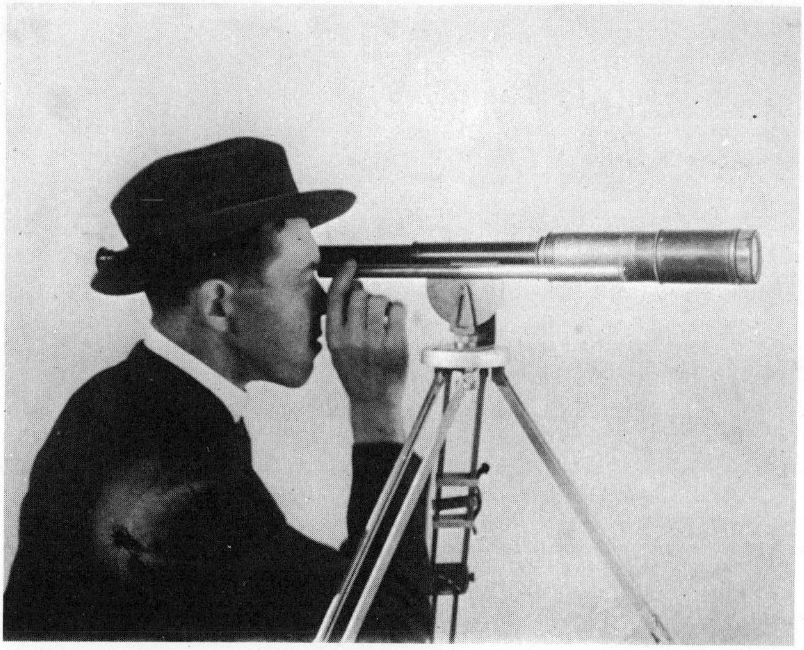

Oscar S. Marshall and his $11.00 refractor telescope, sometimes called a spyglass.

journeyman machinist en route to the stars

Stellafane to Palomar

The Autobiography
of
OSCAR SETH MARSHALL

Edited by
Eva Marshall Douglas

Wm. S. Sullwold Publishing, Inc.
Taunton, Massachusetts

LCC NO. 78-64614
ISBN 0-88492-025-9

First Edition

Printed in the United States

Preface

Journeyman Machinist En Route to the Stars is the autobiography of my father Oscar Marshall, journeyman machinist, artist, writer, photographer, telescope builder. He was born in 1870 and died in 1953. During the last twenty of his eighty-three years he worked on his life story, but did not finish editing it for publication as he had hoped to do. He called it *The Last Journeyman Machinist,* a title that — to those interested in the manuscript — seemed restrictive when compared with the scope of the man's interests and experience. The perspective of the years revealed a progression of right choices that he had unwittingly made. They led this journeyman machinist, known as Oscar or OSM, from toy-shop work and precision-tool making in Vermont to an assignment on the giant telescope project in California. Therefore the title was changed to *Journeyman Machinist En Route to the Stars.*

My father's files included not only authorizations to use reference material and illustrations, but also many of his own photographs that bear the mark of professional work. The Notes contain details from the original manuscript that I felt should be preserved, and other information that supplements various aspects of the account.

I wish to acknowledge my appreciation of those who gave help and encouragement through the years as I made this revision of my father's manuscript.

Among the relatives are two other daughters: Candace and her husband Joseph B. Hartshorn, Beulah and her husband Loyal S. Wright; grandchildren Kathryn D. Smith (Mrs. Harry H.), Paul

M. Douglas, John E. Douglas, Joyce H. Ferrick (Mrs. Albert), Janice H. Jones (Mrs. William), Carolyn W. Duprey (Mrs. Wray), Mary Jane W. Chalmers (Mrs. Thomas); nieces Isabell M. Howard (Mrs. Phillip J.), Virginia M. Dugan (Mrs. William E.), Ruth M. Bjornson (Mrs. Henry), Beatrice and her husband Dana Olden.

Close friends whose association with the family dates back to the early 1920's include Louise P. Seale (Mrs. C. A.), Dorothy Henry, Adaline P. Hagaman, and Elizabeth Hageman. Their counsel and professional advice have been invaluable.

Among other individuals who assisted me in a variety of ways are the following: Mr. Shirley I. Gale, former guest lecturer at the Hayden Planetarium in New York City, New York; my associates in the Writer's Group, Winfield, Kansas; Eugenia M. Savage who not only read the manuscript with a discerning eye, but she also enabled me to see The Hale Telescope at Palomar, California; Dorothy Miers (Mrs. Hubert H.) of Penney Farms, Florida; Mr. and Mrs. Robert Y. Hoeh of Belmont, Vermont, and Penney Farms, Florida.

Ernest V. Flanders of Springfield, Vermont, and contributor of the Introduction was a close friend of my father and a machine-shop associate. He is a member of the Springfield Telescope Makers organization, and, I believe, is the only living member of that group who knew him. I am indebted to Mr. Flanders for the personal touch with which he introduces my father to the reader. I am also indebted to him and to Mrs. Flanders for their extensive photographic work that has made additional prints available for use in this publication.

My husband Elmer H. Douglas — himself a writer — has freely given cf his time and professional acumen. Without his reassurance and his scholarly assistance I would not have tried to edit my father's autobiography.

Finally, I acknowledge with gratitude the personal considera-tion that the publisher Mr. William S. Sullwold has given toward the completion of this book.

Eva Marshall Douglas
Penney Farms
Florida
1978

Introduction

When I learned that the autobiography of my old friend the late Oscar Marshall was being prepared for publication through the efforts of his daughters Eva Douglas, Candace Hartshorn, and Beulah Wright, I was greatly pleased. Oscar and I had been long-time friends in spite of the difference of more than two decades in our ages.

As I read the manuscript in preparation, I noted the attention that Oscar gave to the status of the journeyman machinist. He was one of that disappearing group of men who had earned their title by thorough training in the old-time machine shop. I, too, carry this title of journeyman machinist with a diploma granted in 1916, twenty years after Oscar received his in the year 1896.

Soon after I settled in Springfield, Vermont, October 1, 1920, I became aware of a tall man barely fifty years old working in the Special Tool Department at Jones and Lamson Machine Company. He was the foreman, Oscar Marshall, who held one of the most demanding supervisory positions in the machine shop. Because the men were under pressure to produce accurate results with as much speed as possible, tool jobs seldom had to be re-done.

Within a few months Oscar and I were good friends. A member of the Springfield Telescope Makers organization, he put me in touch with its founder Russell Porter who persuaded me also to join the group. This interest in astronomy and a mutual involvement in local activities drew Oscar and me into a life-long friendship that I have valued throughout the years.

Both of us liked to walk and frequently we set out early

Sunday morning, summer or winter, for a hike into the country. We would head for a secluded spot in the woods, build a fire and cook our breakfast of bacon and eggs with coffee. Usually we managed to return home in time for church services.

In 1923, I had my first Ford. This made it possible to take weekend trips, sometimes to the country home of Oscar's brother George, in Mendon, Vermont. There we would leave the automobile and join other friends in a climb to the peaks of Mt. Killington or Mt. Pico. We felt that the thrill of the trail and the view from the summit were always worthwhile. We explored other Vermont areas, but because of our love for the mountain scenery, we favored the Green Mountain excursions.

From Oscar I derived a life-long interest in photography. I recall his enthusiasm for Mt. Ascutney, which we climbed many times. He was fascinated by that mountain as a subject for his photographic skills. He sought a variety of vantage points from which to capture its majestic beauty and looked for the precise foreground that would enhance that beauty. Oscar was never content with any but the most carefully planned compositions for his photographs.

The same can be said of Oscar's writing. As secretary of the Springfield Telescope Makers organization his style was unique. He loved to write and he sought constantly a better way to express his thoughts.

To sum up Oscar's life let it be said that he was always in search of the best, whether in his work at the shop, in pursuit of his varied interests, or in the cultivation of his relationships with people. He was a man who built lasting friendships. His whole life had unknowingly been a preparation for that great opportunity, which he received when he was sixty years old, to work in the field of astronomy at the California Institute of Technology.

I wish to thank Oscar Marshall's daughters for opening the door so that we might share the innermost thoughts of this man as he struggled to make the most of his life and to achieve his singular success. His story should be an inspiration to all of us.

Ernest V. Flanders
Springfield, Vermont
October 23, 1976

FOREWORD

The reader will perceive that the author did not become a journeyman machinist to the exclusion of other desires and ambitions. He wishes, however, to state without reservation that no group of workers commands his respect and deep admiration as does that great army of journeymen machinists to which he himself belongs.

For those who wish to read about machines and their operations, they may turn to books and periodicals that offer technical treatment of these subjects.

This author is sensitive to the modesty of those whose occupations are not in the field of science. They are curious about scientific discoveries, yet they hesitate to expose their ignorance. Not a scientist himself, the author has been fortunate in his association with men of science. He has found them approachable and has observed their glow of satisfaction when recognized and questioned. One illustration is the story of a humble man of science, Russell W. Porter, whose genius gave birth to the amateur telescope movement. Porter's enthusiasm inspired a group of journeyman machinists in Springfield, Vermont, and they became The Springfield Telescope Makers, builders of what is popularly known as "the poor man's telescope."

In the present account, the author gives the human-interest side of his own story as a journeyman machinist whose machine-shop days covered the turn of the century through the recent completion of The Hale Telescope at Mount Palomar.

Oscar S. Marshall
August 1, 1948
Pasadena, California

Table of Contents

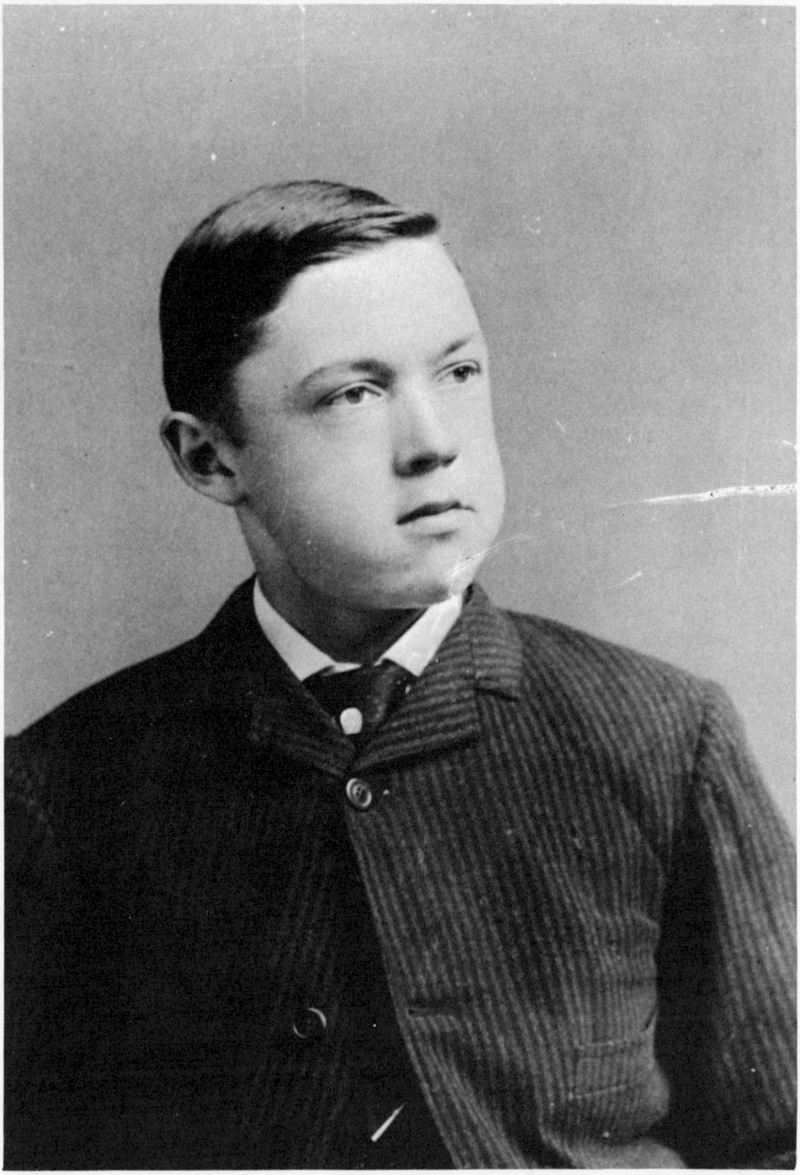

Oscar Seth Marshall

CHAPTER I

DECISIONS

There had been no journeyman machinist on either side of my family. I became one by a kind of necessary choice.

The term journey derives from the French *journee* that indicates a day's length of time. Originally, the word referred to one who worked for a day's wage. Later it came to designate the person who served his apprenticeship in a particular trade and then worked at it wherever he could find a job.

While in training, the journeyman machinist learned to make, to operate and to repair machines and machine tools. He was a familiar character who wore blue overalls and jumper, and carried a ball peen hammer and a 12" ruler ready for use.

Today this tradesman, all but extinct, is no longer the conspicuous individual he became prior to the 1900's. Mass production has made him comparable to a rare book with subject matter of quality that may on occasion answer a particular need.

I was the third of four sons born in a Methodist parsonage to Candace Knight and Alden Velona Marshall.[1] I arrived before daylight on Sunday morning, January 16, 1870, in Cornwall, Vermont.

As a young man, Father had migrated from New Hampshire to Vermont where he met and married Candace Celina Knight of Shrewsbury. While working in a sawmill, he lost part of his right forefinger. Because of this he was rejected as a volunteer for the Union Army. He and his brother Perry decided to prepare for the field of medicine as well as for the ministry. Strict qualifications did not confront young aspirants as they do today. Both brothers fulfilled the requirements of their chosen professions.

Alden Velona Marshall

Candace Knight Marshall

Father and his brother were diligent readers. When Darwin's theories began to appear in current publications, they read his scientific discussions with profound interest. As a result they resigned from the ministry in the spring of 1886 to give full time to the practice of medicine.

Father decided to go to Mayetta, Kansas, where Mother's sister and husband were struggling to make a living on land that had suffered a succession of droughts. Because of limited medical care available, Father hoped to serve the scattered homesteaders in Kansas. He set out alone from Vermont in May, 1886, having planned for the family to follow him. We four boys and Mother remained with his brother Andrew, a farmer who lived in Mechanicsville, Vermont. It was a small village south of Rutland and is known today as Belmont.

The first word my family received, after Father bade us farewell, was a telegram stating that he was dead. A cold contracted en route had developed into pneumonia. When he arrived at our aunt's home in Mayetta, he was beyond recovery. Although buried in Mayetta, his body was afterward removed to the family lot in the cemetery of Wallingford, Vermont.

Seven years elapsed before I made the decision to become a journeyman machinist. During those years I worked in the shop of the Chase Toy Works in Mechanicsville and later transferred to Rutland. I learned how to use the forge and to manipulate simple machines without mangling my fingers. In fact, I became familiar with most of the operations, from measuring a carload of lumber to finishing a product with ornamental scrolls, paint and varnish, and then packing it for shipment.

Dr. George G. Marshall — World War I Chief of Head Surgery, Department Naval Hospital, Ancon, Panama Canal Zone. Retired with rank of Major.

Our eldest brother Velona worked on Uncle Andrew's farm before he entered medical school. George our second brother and I worked at the toy shop. We helped to keep our youngest brother Eaton in school. He was financing his own way through the Methodist Troy Conference Academy in Poultney, Vermont. He joined the New England Conference of the Methodist Church and thereafter served pastorates in Massachusetts for over fifty years.[2]

George had set his sights on a medical career. Four years after our father's death, he entered the University of Vermont Medical College. At the end of the three-year program in 1893, he graduated and set up his practice in Wallingford, Vermont.[3] Subsequently, he moved to Rutland where he established his carreer as an eye, ear, nose and throat specialist. He retired in 1948.

Velona[4], who received his medical degree from the University of Vermont Medical College, began his practice in Moriah, New York, where Father had once held a pastorate[5]. In February 1912, after bringing several patients through severe cases of pneumonia, Velona himself succumbed at the age of forty-eight.

Eaton B. Marshall, Oscar S. Marshall and George G. Marshall in Belmont, Vermont, 1912.

My adolescent years were marked by a certain procrastination mingled with a failure to sense either the merits of life or its punishments. Finally, at twenty-three, I decided to become a journeyman machinist. I made the decision in the early 1890's during the Cleveland administration, when business was slow throughout the country.

While my brothers were preparing for their professional careers, Mother and I made our home in Rutland. The Chase Toy Shop,

Oscar S. Marshall, Eaton B. Marshall, Velona A. Marshall and George G. Marshall taken prior to February, 1912.

OSM photograph of Jackson Pond, now Star Lake, Belmont, Vt. Shrewsbury Peak seen in reflection. Circa 1930.

Employees of Chase Toy Works, founded by Phillip Chase in Mechanicsville, Vt., circa 1870. Courtesy, James Holden, Curator of Museum, Belmont, Vt.

where I was working, failed in 1893. Frank Patch, who was one of my senior associates at the toy shop, explored various places for work. He found a job in Springfield, Vermont. It was thanks to Patch that I made a special trip to that obscure little town fifty miles south of Rutland.

I paid a visit to the small firm of Gilman and Son situated in the heart of Springfield. They were builders of lathe machines designed to turn shoe lasts, gunstocks and hatblocks. Notwithstanding slack business conditions elsewhere in that community, the Gilman Machine Shop continued a lively production in its line.

Wilbert Gilman, the son, was absent at the time of my interview with his father, Farley B. Gilman, who retained major control over the business. My twenty-three years and six-foot height favorably impressed the elder Gilman. I would be a good hand when it came to loading machines on the freight wagon for shipment. Mr. Gilman made no reference to my educational background. He knew that I was the son of a Methodist preacher and probably didn't use tobacco or much profanity, details he had learned from my friend Patch.

Satisfied with my credentials, Mr. Gilman senior enrolled me to begin the three-year apprenticeship that fall.

The achievements of Farley Gilman, founder and owner of the Gilman Machine Shop, merit the consideration given them in the following chapter.

Photograph of R. W. Porter's pencil drawing: Old Mill, near Chester, Vt.

Chapter II

PRECISION-MAN GILMAN

Farley B. Gilman was the founder and manager of the Gilman and Son Machine Shop, where a journeyman machinist received excellent basic training. Gilman had served his own apprenticeship with Parks and Woolson Machine Company. Located in Springfield, Vermont, they were builders of woolen-cloth finishing machinery. This industrial concern, which began before the Civil War, had prospered through the years because of its unique inventions and its able business management.

Mr. Gilman was born December 14, 1834, in Unity, New Hampshire, about twenty miles east of Springfield. Lempster, birthplace of my father, and Unity were adjoining towns. Today it would be possible for two youths, born in adjacent communities, to get into school scrimages. It is doubtful that these men ever met even in their boyhood days.

During his apprenticeship at the Parks and Woolson Machine Company, Gilman discovered the advantages of Springfield's abundant hydraulic power and its fine achievements in the field of mechanics. Soon after he completed his training as a machinist, the young man set up business for himself. In 1853, he obtained a power site at little expense on the Black River, directly opposite the Parks and Woolson Machine Company. There he constructed his first machine shop. He made it known that he would build or repair such equipment as farmers or sawmill owners might need. Gilman was a natural mechanic, well trained for his trade. His ability was recognized and his shop became well patronized.

One day Gilman received a letter from a company in Canada. The writer inquired whether Gilman could replace the broken part of a shoe last lathe. He replied that he could. Little did he dream

Farley B. Gilman of Gilman and Son, taken in 1896.

Gilman & Son employees, 1894, Springfield, Vermont.

that the fulfillment of this request would be the forerunner of his real business, one in which his men would become skilled journeymen machinists.

To do the job required in the request made by the Canadian company, Gilman did some research on machines designed for turning shoe lasts. He located the essential facts that had led to patents

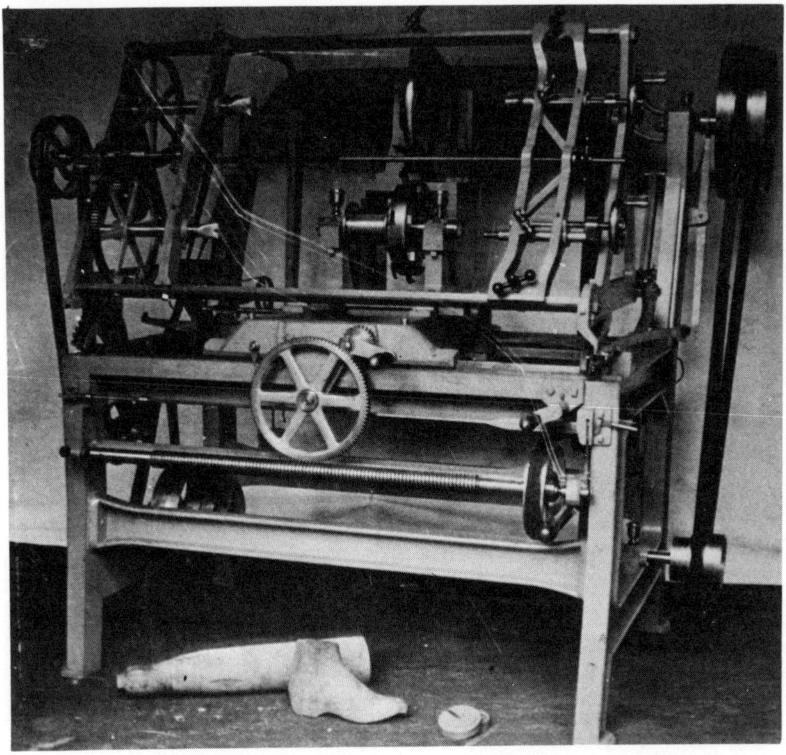

Gilman & Son Shoe Last Lathe from 1890's.

of inventions by Thomas Blanchard issued in 1819-1820. These were special machines made to turn irregular forms for both shoe lasts and gunstocks. It is conjectural which of the two served the greater demand in those early decades of the nineteenth century.

To finance the building of last lathe machines, Gilman needed more capital than he possessed. His wife came from the well-to-do Townshend family. Young Gilman approached her brother F.V.A. Townshend about the enterprise that he had in mind. He persuaded Townshend to contribute, as his share in the partnership, the necessary funds to match his own inventive skill as mechanic. This was in 1861.

The young men began their plans to design and to build a machine that would turn shoe lasts. When they had nearly completed their project and were about to consider the matter of sales, they learned to their dismay that the Blanchard patents, which had

Portrait of Thomas Blanchard from a wood cut made in 1839. Taken originally from Howe's "Emminent Mechanics".

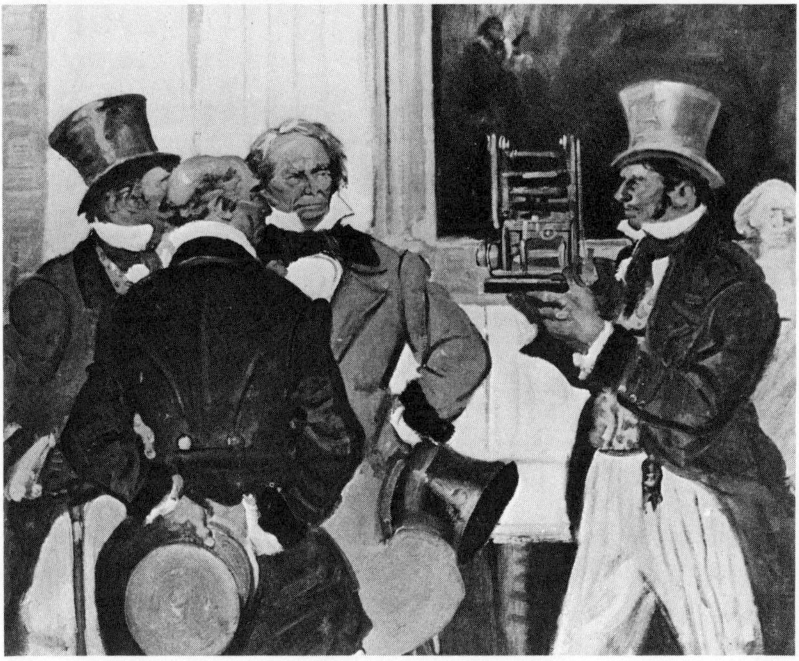

Painting of Thomas Blanchard showing model of his shoe last turning lathe to Clay, Calhoun and Webster. Photograph courtesy S.K.F. Industries, Philadelphia, Pennsylvania.

Blanchard's Gun Stock Lathe, patented 1818. The original of this photograph is in the U.S. Arsenal in Springfield, Mass.

been renewed, would continue in effect until 1870. Gilman and his brother-in-law set aside their invention to wait for that expiration date. They discovered, however, that they could obtain a patent on their own re-designed machine. We shall see that they succeeded in doing just that.

Thomas Blanchard was born June 24, 1788, in Sussex, Massachusetts. Very little is known of that unique inventor compared with the knowledge about the famous Eli Whitney, born twenty years earlier in nearby Westboro. Encyclopedias of the late 1800's make no mention of Blanchard, whereas the 1902 edition of the American International Encyclopedia does refer to him. The significance of his particular inventions merits this belated recognition.

A painting by Benton Clark (reproduced by courtesy of SKF Industries, Philadelphia, Pennsylvania) shows Blanchard as he demonstrates a model of his shoe last lathe to Messrs Clay, Calhoun and Webster, who favored the inventor. It is reported that American jurist Rufus Choate objected to a renewal of Blanchard's patent, saying, "He 'turned the heads' of Congressmen." Blanchard stood by with a model of his machine in hand, while the Congressmen looked in amazement at busts of themselves turned by a copying lathe. (For reference see *Fortune Magazine*, June 1946.)

A Patent Office drawing of the Blanchard shoe last lathe, dated 1819, may be compared with a photograph of a Gilman last lathe built in 1895. Blanchard's patent as incorporated by Gilman in his own shoe last lathe covered a frame, carriage or slide, a swing that supported the model, and a block that was to be turned. Beyond these features, no similarity existed between the two machines.

Blanchard's lathe, made primarily of wood, was turned by hand power. Note the crank at the end of the machine. One observer jokingly remarked that the power was derived from an "armstrong motor." The operator would indeed need muscle and a stout heart plus a willingness to man the device.

A shoe last lathe produced by Gilman and Son, Springfield, Vt., in the 1890's.

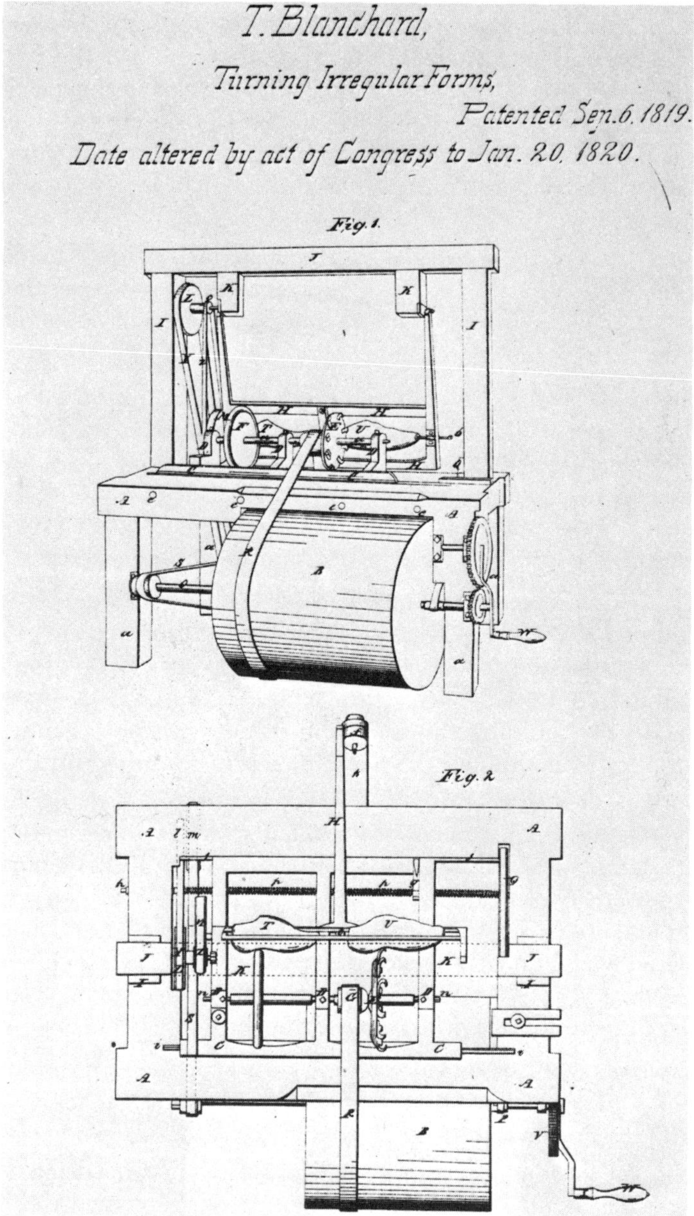

T. Blanchard Lathe for "Turning Irregular Forms" patented September 6, 1819. Date altered by act of Congress to Jan. 20, 1820.

Blanchard's machine required a model for each specific size and shape as well as one each for the right-foot and for the left-foot shoes. These came into vogue by the 1850's It was in this particular area that Gilman revealed his ingenuity. His improved invention made it possible to turn not only right-foot and left-foot shoe lasts from the same pattern, but also ten different sizes from a single model.

As I trained to become a journeyman machinist, I helped to build this Gilman Last Lathe. Build, not manufacture, is the correct word. The term manufacture suggested mass production and the Gilmans were not partisans of that growing trend.

Before I began my apprenticeship in 1893, the original firm, founded in the 1850's and known as Gilman and Townshend, had been dissolved. Farley Gilman's son Wilbert F. Gilman had become partner and heir to the business. Wilbert was a graduate of Boston University. Prior to joining his father's firm, he had taken a two-year course at Worcester Polytechnic Institute in Massachusetts.

The Gilman Machine Shop had been enlarged from its initial size to a two-story structure plus a basement. Power for the machinery was generated by water from the dam constructed by the elder Gilman and controlled by his firm. Farley Gilman had supervised the details for building the shop and its power plant. Considered an authority in hydraulics, he was frequently consulted on the subject.

Located in the basement of the plant were the blacksmith shop, a large lathe for millwright work to handle the pulleys, and the water-power bevel gears. It was also the storage area for supplies of iron and steel. From below, one could hear the merry music of the water wheel and the bevel gears that drove the machinery on the first floor above. Annual inspection of this power plant, penstock and dam was a regular procedure in August.

On the second floor were rooms with machinery, one for work in wood and one for work in metal. A third room was reserved for painting the finished products.

Gilman designed and built an elevator that carried machinery between the basement and second floor. It also lowered the heavy bars of iron and steel to the storage area upon their arrival at the plant..

The shop was air conditioned, so to speak, by movement of the leather belting that drove the main line, counter shafts and individual machines. Electrification of machine shops had been introduced for lighting purposes only. Gilman had installed a generator to meet this need. It was a novel experience to observe the gradual appearance of light inside the glass globes where the bamboo filaments glowed with a warm red color. Their illumination scarcely exceeded that of a good kerosene lamp.

Farley Gilman provided no fire protection beyond his prohibition of smoking at all times. He rarely hired a man who used tobacco. The heating system consisted of the blacksmith forge in the basement, a stove on the second floor, and a large Magee Heater centrally situated on the main floor. Unless regularly cleaned out, this particular stove developed the habit of sneezing a volley of ashes into the atmosphere.

On the main floor, the horse-drawn freight wagons backed into the shop through the wide double doors and the local foundry delivered its castings through the same entrance. It was there that Orin North loaded his one-horse freight wagon with Gilman's crated machines. He transported them to the freight station of the Boston and Main Railroad in Charlestown, New Hampshire. It was six miles south of Springfield and on the opposite side of the Connecticut River. North would say, "It's a good dirt road and in the winter good for sledding."

Such was the setting for the apprenticeship of this journeyman machinist.

Gilman and Son, Incorporated

Springfield, Vt., looking northwest. Photograph by Oscar Marshall, 1920's.

Chapter III

APPRENTICESHIP

Times were difficult in 1893 when Grover Cleveland began his second term as President of the United States. Vermont felt some of the distant rumblings of economic reform for which the country's unemployed were clamoring. In general, the citizens of our small state went patiently about their affairs.

In the fall of that year, Mother and I moved to Springfield, where we decided to make our home for at least three years. Admittedly, we had some doubts about this relatively unknown town. The Johnson's Encyclopedia, which appeared about 1894, made no mention of Springfield, although it did give creditable space to Windsor, Vermont, seventeen miles north on the Connecticut River.

Geographically isolated, the picturesque town of Springfield lay crumpled between hilltops and valleys. From the northwest, the Black River flowed southeast to Springfield and thundered into the chasm that cut through the heart of the town. Energy produced from its falls furnished a dozen water-power sites within a span of four miles. It entered the Connecticut River four miles south of Springfield.

At the center of the town was the incorporated village with its business concerns: machine shops, gristmill, lumber yard and stores. The Adnabrown Hotel and the Spafford Public Library had recently been built. Voters had approved the building of a new high school. The citizens were justly proud of their Volunteer Fire Brigade and its nickel-plated Amoskeag Steam Fire Engine.

Several doctors and two dentists served the community's medical needs. Three churches, the Congregational, the Methodist and the Universalist, stood at respectable distances on Main Street. Only the Universalists lacked horsesheds, never having to pay for their removal when the day of the automobile arrived. A Baptist congregation worshipped in North Springfield. The town of Springfield was on its way to warrant the recognition it receives in the encyclopedias of today.

In the town of Windsor the fortunes of the Jones and Lamson Machine Company, which was located there, had vacillated for over fifty years[1]. When Springfield hovered on the brink of economic uncertainty in 1888, some of its influential citizens pooled their resources to buy that machine-tool company. They transferred it from Windsor to their own inconspicuous town where the Black River waterfalls made hydraulic power readily available. It proved to be a significant venture that brought new life to the threatened area and restored to Springfield its once flourishing economy.

By 1893, however, the national economic slump produced an adverse effect on the business of Jones and Lamson Machine Company. Had this not been the case, I might have found employment with them instead of with Gilman and Son.

When I began my apprenticeship, my personal assets in cash amounted to less than three dollars. Our house rent was $8 a month. Coal sold for about $7 a ton and good hardwood for $3 a cord ready for the stove. Milk was four cents a quart. Butter sold for twenty cents a pound and eggs for ten to twelve cents a dozen. Potatoes were thirty to forty cents a bushel and apples $2 a barrel. Around Thanksgiving one could put down a side of pork or a quarter of beef for a moderate sum. Clothes and footwear were equally reasonable. In spite of the depression, Mother and I managed to get along. We made ends meet.

The first year as an apprentice at Gilman and Son's my daily wage was seventy-five cents. The second year I received ninety cents a day and the third year $1.10 was my daily wage. Monday through Friday were ten-hour workdays. On Saturday we worked nine hours plus a few extra minutes to sweep the floors.

The work force comprised sixteen men including the two Gilmans who also carried responsibility for office details. All the men were seasoned mechanics in their fifties, all except Eugene Lockwood and myself. Eugene, in his final year of apprenticeship, became one of my close friends. Elliot Thompson, a real old-timer, had learned his trade at Gilman's and never worked elsewhere.

The output of our work force totaled three or four last lathe machines a month, each one priced at $500 f.o.b. in Springfield, Vermont.

Buotelle was Gilman's right-hand man and his most highly paid workman. He received $2.75 a day. He was the only workman assigned the responsibility of making the cutter-head spindle that turned at 4,000 rpm's. Occasionally Boutelle would have a problem with one that overheated when running at full speed. Years later the cause of his difficulty was discovered. He had used a fine grade emory with lead lap and oil to produce on the bearings a polish that he was unable to obtain with a dead smooth file. The metal, not treated for heat, became charged by friction with the emory. The science of metallurgy had not at that time become a recognized factor in the manufacture of machinery. Nor were alloyed castings and high-speed steel on the market. Norway iron, some Bessemer steel from England, and brass were the principal metals used by the Gilmans. On occasion, while a shaft ran at filing speed, one of them would feel a bearing to detect the presence of any ridges. These would be due in some cases to the nature of the iron and not necessarily to poor lathe work. Bessemer steel was generally free from such defects.

The Gilman Last Lathe had acquired renown for its snappiness, versatility, accuracy and reasonable durability. Customers were willing to wait a year for their orders to be filled by Gilman products.

One foreign customer turned out to be completely dissatisfied. Gilman and Son had a healthy European business with agents in both England and Germany. One of them sold a machine to a man in Armenia. When the machine was set up and in operation, Armenian authorities forbade its use. They ordered it to be removed from the country. They believed it to be the work of the devil. Their reaction appeared to derive from the high speed of the cutter-head spindle with its 4,000 rpm's. The noise was too ominous for the authorities to tolerate.

The Gilmans, hesitant to indulge in anything suggestive of mass production, eschewed the encroachment of modern ideas on their product. They bought practically no equipment.[2] The elder Gilman had made many tools for shop use. Most of those required for building the last lathe had originated right there in his plant. He encouraged his apprentices in the very beginning to make their own calipers.

All of our work was done by rule of thumb, without blue prints. None was ever seen in shop or office. One of the Gilmans explained or demonstrated every procedure to us. For each assignment our goal was to achieve the accurate fit of a particular part into a particular piece of machinery.

When I appeared on the scene, the younger Gilman was beginning to assume considerable oversight of the shop program, which included training of the apprentices and general supervision of all work in the shop. Wilbert Gilman's mechanical skill did not equal that of his father. Technical training does not necessarily produce a skilled mechanic any more than musical training will make a virtuoso of a performer. Nevertheless, Wilbert had profited by his formal studies[3] as well as by the expert example of his father.

The Gilmans tested their products before they were considered ready for shipment. Wilbert or his father would turn a few sample last lathes to detect the slightest flaw or defect. Testing a lathe was an event in shop life when a man's work was thoroughly examined. If an error was discovered, it received careful scrutiny to determine whether it was due to a simple mistake or to an attempt to conceal bad workmanship. If the latter, it meant a grueling session in the Gilman office, followed by needling of shopmates.

Young Lockwood, who bubbled with enthusiasm much of the time, was caught on one occasion when he tried to disguise an inferior job. The elder Gilman, fortunately, made the discovery. He had mellowed through the years of experience with his men in the shop. Yet he viewed Lockwood's predicament with less than his usual calm. The defect in workmanship was serious enough, but to Farley Gilman it was reprehensible that a young man of Lockwood's reputation for integrity should have yielded to the temptation to hide his mistake.

Lockwood's problem lay in the adjustment of a small taper pivot that he tried to use. He had spent considerable time making that pivot and, in a moment of weakness, it occured to him that by slightly bending it he could solve his problem. The pivot, however, failed to function accurately and the subterfuge was discovered.

This incident is not mentioned to reflect unfavorably on the young man. Gene Lockwood possessed more than ordinary talent for shop work and he became conspicuously dependable. Nor do I intend to throw other than creditable light upon the jealous oversight of the Gilmans. They sought to maintain the high reputation of their machine products and of their training of journeymen machinists.

My relationship with the Gilmans was on the whole very satisfactory. Nevertheless, I did run into one most embarrassing situation. It came about through what we called the oil-can or monkey-wrench trick by which I ruined the main shaft for a last lathe. The younger Gilman had tutored me in making a shaft with acceptable results. "We'll clinch the lesson now by making one more shaft for another machine nearing completion," he suggested.

I shall not elaborate on my misfortunes, but I had placed the oil can midway on the track of the lathe and forgot about it. The continued advance of the carriage on the track pushed the oil can along until it was forced into the shaft and cut a nasty groove.

It was inconceivable to send out a piece of work with a scar on it. I was properly exercised over this first real casuality of my apprenticeship, but apologies were of no avail with my instructor who at times had difficulty in controlling his temper. My mistake humiliated me more than his efforts at verbal chastisement. I was ordered to start work on another shaft.

Having failed to regain presence of mind from the chagrin of my mistake, I repeated the same blunder at about the same place. And I was in for another session with young Gilman, barely four years my senior. By this time the old gentleman had become aware that something was amiss between his son and me. He edged his way into the situation and suggested that he take over. With a certain quiet kindness he said, "Take the bar upstairs. We'll be able to use it someday. Go down to the basement and cut off another piece. We'll not spoil this shaft."

With the third shaft satisfactorily completed, the elder Gilman worked with me until we had it placed in the machine for which it was designed. He did the spotting of the shaft where the set-screws of the drum and the drive-pulley came in contact. He did this with a bastard file, saying as he did so, "That's good enough."

It so happened that the unit had to be taken apart during the testing of the lathe. Wilbert was checking the machine. He noticed the rough character of the spotting. He took me to task for the work saying that he supposed I could do a better job. I replied, "Yes, I think I might, but your father did it."

Within the shop, especially during the work hours, a paternal spirit prevailed on the part of the Gilmans. One day at closing time, young Lockwood left by the usual door with a leap and a dash that overturned Wilbert's stool. Wilbert called to the young man and demanded that he return, set the stool straight, and then leave in a proper manner. Much to my surprise Lockwood obeyed. On another occasion Wilbert said to him, with a bit of exasperation in his voice, "Young man, you've just about reached the end of your rope!"

"When I do, I'll bust!" he rejoined.

These episodes were merely part of the ups and downs of the old-time machine shop.

The course of our apprenticeship did not follow a strict plan, but was adapted to work conditions and to business demands. In the process, it afforded the apprentice firsthand experience and excellent training in every phase of shop work as it existed at the time. We gave particular attention to handwork, such as the accurate fitting of small parts with square holes onto square bars. Today, with the development of mass production in the automobile industry, machinists use a broaching process for these jobs and a variety of other meticulous fittings.

The apprentice who served with Gilman and Son sought to avoid hearing them say, "You'd be a good one to file cannon balls."

CHAPTER IV

DIVERSIONS AND A DETOUR

In the summer of 1895, during my apprenticeship at the Gilman Machine Shop, Wilbert Gilman took six of us fellows on a wagon trip. We drove northwest to Snake Mountain and crossed Lake Champlain to visit iron ore mines in eastern New York State.

On the return journey we waited at Port Henry, New York, for the side-wheeler steamer "Vermont", scheduled to stop there on its way to Burlington, Vermont. Two hours late, Captain Rockwell was not cheered by the sight of our Watertown Wagon with its pair of horses waiting at the dock.

The gangplank was shot to the wharf. Captain Rockwell gave orders for us to drive on board and for the steamer to pull away. Both horses reached main deck, but only the wagon's front wheels had made it, with the canopy-top still wedged between the main and upper decks. The rest of the vehicle hung in cantilever fashion over the edge of the vessel.

"Take off the front wheels," commanded the captain. Off they came with wrenches that appeared from somewhere. Down went the front end of the wagon and up rose the gangplank. Our cargo was on board and we relaxed for the crossing to Burlington.

On the lakeshore of that city we set up camp for the weekend. We planned to hike to the top of Mount Mansfield, an altitude of 4,500 feet, highest point in Vermont. We reached the summit of the mountain's giant profile to find the wind blowing a terrific gale. A couple of us approached the brink of a precipice before we discovered the danger, but we made it safely down the mountain to our horses and carriage.

From that adventure we drove southeast to Barre, Vermont, there to visit its granite quarries. By the weekend we arrived home with a collection of stories for our families and friends.

In August of the following summer, I completed my three-year training at the Gilman shop. Gilman and Son presented me a bonus of $100.00 with a letter saying, in substance, that Oscar S. Marshall had satisfactorily achieved his three-year program and was now a certified journeyman machinist. The Gilman's requested that I continue in their employ at $2.00 a day. I accepted their offer.

Before I began my technical career as a journeyman machinist, Mother and I took a week's vacation trip with horse and carriage. We drove to a small community near Stratton Mountain in southern Vermont where her father and her brother Charles Knight lived. My eighty-year old grandfather was straight as an arrow, a handsome man with snow-white hair and rosy cheeks.

In the village of Stratton stood a small white schoolhouse where a certain young lady, as yet unknown to me, would one day be a schoolteacher. Her story will come later.

My career as a journeyman machinist began in the fall of 1896 at the Gilman and Son Machine Shop. The Gilmans, reluctant to consider new methods or plans for expansion, continued their old-time procedures. It is to their credit that, in spite of this, they maintained a healthy business.

One event I recall drove the Gilmans to put pressure on all of us in the shop. The Empire Last Company of New York had suffered a total loss by fire and had rushed one of their agents to Springfield. He counted on the Gilmans to recoup the situation for his company. In an incredibly short time we had finished every last on the floor of the shop. Each one was crated and labeled for The Empire Last Company, New York, New York. When the final machine was on the freight wagon, the agent climbed aboard, clapped his coattails in triumph and waved adieu to the Gilman force.

To say that I saw my role as a journeyman machinist to be threatened at this particular time by the progress of mass production would be inaccurate. Nevertheless, stirrings within me did divert my attention from the daily routine of the job at the shop. Instead of mulling over shop-related problems, I became interested in social and economic questions, matters of religion and philosophy. I found intellectual stimulus in the thoughts of Herbert Spencer and

William Lecky, in articles of *Harper's Magazine, Scribner's Magazine* and *American Review of Reviews.* The writings of Sir Arthur Balfour ploughed up my ripening mind as did the works of the educator and naturalist David Starr Jordan. I began to keep a notebook of arresting quotations that also included lines from certain poets, Lowell, Tennyson, and Shakespeare, in particular.

Persistent disenchantment with the job led me still further afield.[1] I read something about James Roscoe Day, Chancellor of Syracuse University, that prompted me to write to him. I inquired whether he would accept me, at twenty-seven years of age, as a special student at the University. He wrote me a most encouraging reply and suggested that I make plans to come to the University.

Brother George, who was established in his medical profession in Wallingford, Vermont, agreed to make a home for Mother while I went back to school.

In the fall of 1897, I entered the College of Liberal Arts of Syracuse University. The Dean of the College made two objections to my entrance there: a man of my age should be a producer, and special students did not increase the number of graduates listed in the catalogue. A note, however, from the Chancellor settled the matter. The Dean made out a schedule of subjects that were of most interest to me. These included ancient, medieval and modern history, sociology and economics, philisophy, psychology, biology, English composition, and public speaking. As an optional lecture I chose astronomy, the significance of which will appear later in this account.

My last days in the classroom had been spent in a little red schoolhouse just before I turned sixteen. Now, at twenty-seven, I felt like a man from the dead moon. It was nip and tuck for me to keep up with the classes. I decided that, if light didn't break by December, I would abandon my dream of intellectual improvement at the University.

The fall term began with what the students called a salt rush between the freshmen and the sophomores the first day after the chapel service. To avoid confusion, the juniors marked the cheeks of the freshmen with a red cross. The sophomores, who had reserves of ammunition, opened fire when we emerged from the chapel. Six feet tall, I was among the first to receive a volley of salt. We gathered our forces to rush the enemy. Outnumbering the sophomores, we were not long in cleaning the field, or so we thought. We marched around it four-abreast, then three, two and finally single file. Suddenly a stalwart sophomore seized me and, ere I was aware of him, he had me down for a salting. He had not gone far in the pickling process when several freshmen came to my rescue. Together we gripped that sophomore by all fours and heaved him down the fifteen-foot slope below the campus level. I should add that later in the year this particular sophomore and I became good friends.

When various tests were given in December, I managed to meet the academic standards. Instead of going home for the Christmas holidays, I remained in the city to complete reading assignments and written work. In the mid-year and final examinations I met with encouraging success.

Through the summer vacation of 1898, I worked at the Gilman Machine Shop. In the fall, when I resumed my studies, I added a course in first-year German for which my roommate coached me to advantage.

As a partial fulfillment of required courses, students in the fields of economics and sociology visited certain industries and institutions. I was astonished to see the mechanics (called instrument workers today) wearing white shirts and neckties, and working at benches as clean as a cook's kneading board. We also visited the Institution for Feeble-Minded Children, 85% of whom were said to be the offspring of alcoholic parents.

Early in the fall of 1899, William Jennings Bryan, Democratic Party Leader and unsuccessful candidate for the presidency of the United States, visited Syracuse. He delivered an off-campain address to the packed convention hall. I had heard outstanding speakers — politicians and statesmen, religious leaders, reformers and educators. Yet in the command of English, Bryan seemed to excel them all. When he concluded his address, hundreds of people filed by his podium to shake the hand of the celebrated commoner. Although I may well have been the 500th person in line, I had never had my hand grasped for such a hard handshake as that of Mr. Bryan. I was thrilled to have heard the man speak, but my confidence in his judgment had been shaken in the process.

During the McKinley-Roosevelt presidential campaign in 1900, Theodore Roosevelt visited Syracuse. The city's convention hall was too small for the crowd that gathered to hear Teddy Roosevelt, hero of the American troop victory over the Spanish forces at San Juan Hill in 1898. Hundreds of students waited for the hall doors to open. When they did, we were literally carried inside by the mob. Human ladders appeared everywhere along the walls where students and citizens alike climbed to the cross timbers and struts of the building's structure. Pandemonium reigned until Teddy, wearing his Rough Rider's uniform and flashing his characteristic smile, finally reached the speaker's stand.

Sometime before midnight the great hall emptied itself. Hundreds of us found our way to Teddy's hotel and stormed his window until his spectacled eyes and flashing smile gleamed down at us as we yelled, "We want you, Teddy!" He acknowledged our support and with that he lowered the window. We dispersed and returned to our rooms.

Christmas holidays of 1900 were upon us and I decided to spend them in Wallingford. It was the end of the first term of my fourth yeaı at the University. The "Wolverine" carried me to Albany, New York, and the "Green Mountain Flyer" dropped me off at Wallingford, Vermont. Each time I left the central part of New York State for the hills of Vermont, whether summer or winter, I felt that I was returning to my native clime.

A day or so after I reached George's home, where Mother was still living, I received a phone call from my former employer Wilbert Gilman, in Springfield, Vermont. He asked me to pay him a visit and to do so at his expense.

When I arrived at his office, Gilman made a few incidental remarks, and then went straight to the reason for which he had asked to see me. His superintendent and right-hand man had developed cancer. Would I consider coming back to take his place at Gilman's Machine Shop?

With very mixed emotions and grave doubts I went back to Wallingford. I had promised Mr. Gilman to give his request serious consideration. Words of the Dean at the University rang faintly in my ears, "A man of your age should be a producer."

With the family I weighed the pros and cons of the total situation. Finally I called Wilbert Gilman to say that I would return as his shop superintendent to work under contract until January 1, 1902.

I made a quick trip back to Syracuse, gathered my belongings, said adieu to my professors, then took the night train out of town.

Oscar S. Marshall, circa 1900.

Mabelle Maranville, circa 1900.

CHAPTER V

A FUNERAL AND WEDDING BELLS

Let me go back a bit to the summer of 1899. During that vacation from Syracuse University I had a job at the Gilman Machine Shop. Early one evening, while following the wooded trail in a fisherman's paradise, I came upon a soft pink lady's slipper, also known as Indian moccasin. As I gazed at the fragile beauty of this exquisite woodland orchid, my thoughts turned to another midsummer's day.

George and I had been riding together with horse and buggy, traveling along the Otter Creek Valley between Wallingford and South Wallingford, when we met a milk wagon on its way from the railway station. This was not an uncommon sight, but the dark-eyed maiden nonchalantly driving her pair of horses was more unusual. She barely noticed our passing rig, and two years went by before I saw her again, during that summer of 1899 at a funeral service.

I received word that the wife of Uncle Charlie had died in Ludlow, Vermont, sixteen miles north of Springfield. Taking time out from work, I cycled there for the funeral, where I was pleasantly surprised to see the maid of the milk wagon.

At lunch after the service, I sat opposite that young lady, whose name I learned was Mabelle Maranville. She remained indifferent to the various dishes of food that I carefully passed to her. Later she agreed to go for a ride with several of the younger generation. Before the day was over, we discovered a mutual interest in the family photograph album. Mabelle, eldest daughter of Alice Coleman and Edward Dwight Maranville, promised to send me a picture of the Maranville family and one of herself if she could find it. We made arrangements to correspond in the fall after I returned to my studies at Syracuse University. Mabelle planned to teach in the little white schoolhouse of Stratton, Vermont.

In due time, we made plans for wedding bells to ring March 12, 1902, at the Maranville farm on Crow Hill above Danby, Vermont. It was situated in the Taconic Mountain Range west of the Otter Creek Valley where Vermont winter weather and travel conditions were unpredictable. The day before the wedding a snowstorm blocked the road up the long hill to the farm. We menfolk worked hard to clear a passage through snowdrifts in time for the wedding guests who were expected to arrive the next morning.

Mabelle M. Marshall, circa 1902.

Oscar S. Marshall, circa 1902.

March 12, 1902, dawned clear and cold on Crow Hill. Above the farm, nestled deep in snow, soared a single crow in lonely flight, and inside the farmhouse, a bustle of activity. By nine o'clock travelers were in sight, and not long afterward all guests were accounted for: two ministers including my brother Eaton; two doctors one of whom was my best man, brother George; Mabelle's maid of honor, Olive M. Prouty; and the pianist our friend, Ethel Roberts (Mrs. Myron); plus the sympathizers.

With the first measures of the wedding march, I started down the narrow stairway. All at once I felt something let go. Were the strains of Lohengrin too great? My new President Suspenders had

The farmhouse where Oscar and Mabelle were married March 12, 1902.

failed to live up to their trademark. A loop had given way. I remembered the violinist who continued to play on one unbroken string to the end of his piece. I told myself that if I couldn't make it through the wedding march with one missing loop I would deserve to lose my bride. No one appeared to notice that I was ill at ease. Did anyone suspect that I was trying to imitate the swaggering gait of the ancient and honorable artilleryman? Apparently not.

After the wedding ceremony, when the knot was well tied, everyone sat down at a long table bountifully prepared for the occasion. Mabelle and I had little time to spare before catching our two-o'clock train in the valley. My father-in-law had hitched his pair of horses to a sleigh. In went our luggage and off we rode down the snow-packed trail to Danby Depot some distance below.

At the railroad station, Mabelle introduced me to an acquaintance. This was my first post-wedding shock. "Mr. Hawkins, I want you to meet my husband, Mr. Oscar Marshall." I must have missed that word in the marriage ceremony.

The train pulled out of Danby Depot and we began our brief wedding journey to Springfield, Vermont. Before it was over, I received a second jolt to my sensitive nature. Later that afternoon, Mabelle turned to me and said, indicating a middle-aged gentleman sitting across the aisle, "Oscar, I hate to think of the time when you will look as old as that man." With that shocking intimation of some distant future, we set ourselves to the joys of the present.

Our journey took us through Hoosick Falls, New York; Hoosac Tunnel and Greenfield, Massachusetts; and Bellows Falls, Vermont, to Charlestown, New Hampshire. There we made connections with the Springfield Electric Railway that delivered us to Springfield just before midnight.

We began housekeeping in a small upstairs apartment that I had engaged and equipped. We had been liberally showered with wedding gifts. Our little home on Summer Street became the ren-

Oscar and Mabelle's first home, an apartment upstairs in this house on Summer St. in Springfield, Vermont.

Mabelle Marshall, May, 1902.

dezvous for many happy occasions both church centered and community oriented.

I was happy, for the time being at least, in my new situation as Superintendent at Gilman's Machine Shop. My salary, not wages, understand, was quite a respectable figure for those days. Hence it was not long before Mabelle and I began to think about building a house of our own, which we realized within the next three years.

Left, Fellows Gear Shaper and Black River. Foreground, Parks and Woolson Co., Gilman Shop to right of center. Methodist Church steeple, and Hartness residence to right of that, seen through the trees. Springfield, Vt. 1920's.

Chapter VI

UNEVEN RHYTHMS

Not many days after we had settled in our small apartment in Springfield, Mabelle asked me at breakfast, "How do you like the cereal?"

"Fine," I replied, bound to please. Grape-nuts cereal was one of my favorites.

"Well, Oscar, do you know what you are eating?"

"Sure, it's grape-nuts cereal, isn't it?"

"No," she replied with mischief in her eyes. "It's dried brown bread crumbs with sugar and cream."

"You're a modern version of Eve!" That was all I could say. Packaged foods did not impress Mabelle, who was accustomed to home-grown farm products.

With certain frustrations, Mabelle adjusted to life in the small apartment in town. I continued to face a long-term dissatisfaction with my career as a machinist. Before our wedding I had considered an offer to serve a small pastorate in western Massachusetts. This opportunity did not materialize because my contract with Mr. Gilman at the time ran until January, 1902, before which date he would not release me. When I confirmed my plan not to continue beyond the terms of our contract, Mr. Gilman hired another shop superintendent who assumed his duties in October, 1901.

By December, an explosive situation had built up in the shop to which two elements contributed. In the first place, the new superintendent didn't follow Mr. Gilman's shop methods. In the second place, the irritating manner of Mr. Gilman himself created an increasingly tense atmosphere among the workmen.

One morning Mr. Gilman requested that I meet him in his office. Having sensed the feeling of unrest among his men, he asked me what I would do in his place. As I did not discern what lay behind his thinking, I had nothing in particular to say. He may have hoped I would propose something to ease the situation. He left for home about ten that morning.

Within the next two hours it became more than likely that the shop force would quit en masse, except for Mr. Thompson, the oldest man on the floor. He had worked there since he served his apprenticeship under the elder Gilman, now deceased. Each one of the workmen knew that it would be easy to find employment in one of Springfield's other machine shops.

I had no real authority over these men, although they still looked to me for guidance when problems arose. During those hectic two hours before midday, I cannot say that any concrete solution evolved in my mind.

As the noon hour approached, I asked one of the older men, fully determined to bolt the works, if he would tarry with me at the waterwheel. I proposed to close it. I consulted none of the others, but the sight of the two of us in a huddle was sufficient to arouse their curiosity and they all drew near.

With the entire group assembled, I put the question to all of the men, "What is it that we fellows want?"

"A nine-hour day and a shop foreman," said one of the men, and the others assented.

"If we get these concessions will everyone of us remain loyal to Gilman?" They all agreed, every last man of them. I had urged unanimity.

At that point, I seized the opportunity to mention some matters that were relevant to the over-all situation. I could do it impartially because the men understood that I was leaving at the end of the month.

I pointed out the following facts for the men to think about. While our demands were fundamentally sound and just, no one of us had created the business. Each of us had benefited by the opportunity for steady employment that the shop gave us. None of us had the responsibilities and concerns carried by the proprietor, Mr. Gilman. It was within our power but hardly our prerogative to wreck his business by our defection at that moment. I urged that we act with care.

The men considered these comments and finally acknowledged their genuine approval. Unanimously they agreed that I should carry their request to Mr. Gilman. I consented to do so and promised to wait upon him early that afternoon. The men said that they would delay their action until the issue was settled.

Mr. Gilman received me in a friendly spirit. I approached the touchy subject from the riparian-rights viewpoint, the stream of profits likened to a river along the banks of which adjoining interests were recognized.

I had tried to help the men in the shop understand the benefits they enjoyed because they were in Mr. Gilman's employ. Now I wanted him to see how much he in turn had benefited by the work of his men. I pointed out that, whereas the workmen had not been responsible for creating the industry owned and operated by him and his father, their enterprise would not have succeeded without the cooperation of these men. Furthermore, they possessed no inherent property interests in it. They were ignorant of the receipts and the expenditures of the business to which they contributed their skills. As proprietor, he alone had access to those facts. The wages he paid and the hours he set were at his discretion. I also called to his attention the fact that no law now prohibited his dissatisfied men from leaving their jobs to look for work elsewhere. To this he agreed.

As I left him, Mr. Gilman said he would visit the shop before closing time that afternoon to give the men his answer.

The two machine-tool industries in Springfield, Jones and Lamson Machine Company and the Fellows Gear Shaper Company, had voluntarily adopted the nine-hour workday.[1] Mr. Gilman could not afford to risk losing his force of skilled machinists to those machine shops. He grasped the import of the situation and acted accordingly.

Just before five o'clock, Mr. Gilman came into the shop. He asked me to tell his men that their requests would be granted at once and that the shop would close at five o'clock that night instead of at six. Mr. Gilman wore the expression of one who had undergone a real conversion.

The day drew near when I expected to have more time to write, but this anticipation was short-lived. Mr. Thompson informed me that he was to see Mr. Gilman on behalf of all the men in the shop, who had requested that I be retained as their foreman. Perhaps this was a logical outcome of the recent crisis, but it was not in accord with my own desires.

Within a few days, Mr. Gilman invited me to his residence. He presented to me his formal offer of the superintendency of the shop with an annual salary of $1200.00 and a two-weeks paid vacation. I asked for time to consider what was for me an unpromising turn of events.

As I weighed the possibilities, I saw one that favored my remaining at Gilman's. There would be an opportunity in the growing industrial community of Springfield to test some of my ideas about social progress that were restless for expression. I chose to accept Mr. Gilman's offer.

In the discussion of my new responsibilities with Mr. Gilman, I stipulated that he should leave me free to supervise the workmen as well as to hire and to discharge them. If and when he observed a situation in shop matters that he wished altered and that derived from worker-management conditions, I asked that he take up the question with me personally. We would discuss jointly the workmen's wages. Under this agreement I remained four and one half years as superintendent of Mr. Gilman's shop, where I was working when Mabelle and I were married.

The Marshall home at 135 Wall St., Springfield, Vermont, about 1908.

Other developments contributed to uneven rhythms in that period of my life. Mother died July 31, 1903.[2] Mabelle and I planned and built our house. The Young Men's Christian Union was organized in Springfield. I made the decision to leave Gilman's employ in order that I might try a literary venture.

Wall Street with Hartness Park in the background, about 1912. Marshall home is near center, with darker roof. OSM.

In March, 1905, Mabelle and I moved into our new home at 135 Wall Street in Springfield. Located at the east end of the village, it was known as the "yaller house across the gulch." Both the broad front verandah that ran the length of the house and its generous Hawthorn gable overlooked the deep wooded ravine. It was an ideal haven for birds, the redstarts, black and white creepers,

Looking east toward Hartness Park in winter, circa 1912. OSM.

a variety of woodpeckers, red-breasted grosbeaks, vireos and thrushes. The wood thrush nested within twenty feet of the house. We found great joy listening to his evening melodies. Also lingering in memory is the song of the veery as I used to hear it in flight back and forth across the ravine. Mabelle declared that she always knew when the tenth of May arrived, for then she heard the oriole's song bubbling from the top of our rock maple tree.

Oscar and Mabelle's home on Wall Street in 1930. OSM.

It was in this house on Wall Street where our three daughters were born, Eva Theresa in 1906, Candace Alice in 1909, and Beulah Mabelle in 1911. Our yard became one of the neighborhood playgrounds where children's voices called at hide-and-seek late into summer evenings. When Mabelle and I visited Springfield forty years later, we parked in the business section of the town. A young policeman stepped up to our car and introduced himself as Bernard Lashua.[3] He had been one of the neighborhood gang who played in our yard.

Another development during this period was the organization of the Young Men's Christian Union. Blessed with abundant energy in those years, I found a channel for service on behalf of the young men who worked in Springfield's machine shops. Business was good. This meant that the village was growing in numbers. Yet it was remote from city attractions and offered little to the young men for their leisure time.

A group of us started a movement to help meet the situation. We aimed to raise funds sufficient to engage a director and to rent rooms for the use of this new organization, the Young Men's Chris-

Oscar in his "den" at home on Wall Street, early 1920's.

tian Union. Without the farsighted interest of Mr. Charles Woolson, brother of William D. Woolson, our meager beginnings would have been nil compared with the need. Mr. Woolson volunteered to pay half the salary of a secretary-director and the expenses of the rooms on condition that the other half be provided by the organization itself. The life of the Y.M.C.U. was assured. Later it became the Community Club.

A further development, and one with more far-reaching consequences for my life, began in mid-winter, 1906. I had been Gilman's shop superintendent nearly four and a half years. During that time he had not thought it necessary to reconsider or alter the policies that had been those of his father. I grew increasingly restless with the desire to write.

One afternoon in February, 1906, I left the shop in Gilman's charge, saying that Mabelle and I were going for a sleigh ride. We drove out to the farm on the edge of town where her folks now lived. As we rode along, we discussed plans that were already under consideration. If they materialized, I would have my opportunity to write. We agreed that the next morning I should give Mr. Gilman notice of my intention to leave his employment.

The following day I made my announcement to Mr. Gilman. I suggested that he set the date for my leaving to suit his convenience, but that I would fix July 1, 1906, as the limit of my contract with him. He accepted that as satisfactory. (Within a year his health was impaired and he sold his business, which later went to United Shoe Machinery.[4])

When I left Gilman's employ in July, I expected to collaborate with the printer, Mr. Hall, in the creation of a community newspaper to be named *The Town Weekly*. Because of a combination of circumstances, this literary project did not materialize, but another one gave evidence of promise.

I had formulated plans to edit a monthly magazine that I called *The Vermont Review*. In spite of encouragement and of what seemed to be sound business counsel for the undertaking, only two issues came off the press. I faced the grim necessity of discontinuing my cherished enterprise. Not too old to survive both disappointment and some financial losses, I returned to my trade of journeyman machinist. Experientia docet stultos[5] . . .

Oscar S. Marshall on the job at Jones and Lamson Machine Shop, June, 1915.

CHAPTER VII

AT JONES AND LAMSON'S MACHINE SHOP

One day while struggling with my ill-fated *Vermont Review,* I had met Frank Hill[1], Shop Superintendent of the Jones and Lamson Machine Company. When I mentioned my current literary venture, he said, "If it doesn't work out, come to see me." Since it hadn't worked out, I contacted him in the summer of 1907. I found steady employment as a journeyman machinist at the Jones and Lamson Machine Shop. This was the beginning of a series of remarkable relationships and events.

One of my associates, and a close friend at the shop, was Ben Billings. We had worked together in Mechanicsville at the Chase Toy Works. They went into bankruptcy during the depression. Seeking employment, Ben visited Springfield where I was serving my apprenticeship at Gilman and Sons. Jones and Lamson's hired him and Ben became one of their finest salesmen.[2]

The Jones and Lamson Machine Company was a seedbed in which the creative ideas of gifted men had an opportunity to grow. Among these gifted persons were James Hartness, William D. Woolson, Edwin Fellows, and William Leroy Bryant.

Mr. Hartness, later Governor of Vermont, was the shop's first superintendent and eventually the Company Manager. Hartness was an inventor rather than a journeyman machinist. In 1891, with the invention of the Hartness Flat Turret Lathe[3], he established himself as a pacemaker among machine-tool builders. Investors began to take note of Springfield, Vermont.

William D. Woolson, able business executive and talented mechanic, shared with Mr. Hartness the management of the Jones and Lamson Machine Company, and was its treasurer. His gift

Left, James Hartness, about 1920. Courtesy Jones and Lamson Co. Right, William D. Woolson of Jones and Lamson Co.

Left, Benjamin L. Billings. Courtesy H. C. Billings. Right, Superintendent Frank Hill, Jones and Lamson Co. Courtesy E. Johnson.

Left, Edwin R. Fellows about 1926. Courtesy Mrs. Fellows.

W. LeRoy Bryant. Courtesy Mrs. Bryant.

Jones and Lamson Machine Company, 1920.

of discernment helped to insure success when business conditions were uncertain. Both he and Mr. Hartness could detect the worth of an idea and were willing to foster its development.

Young Edwin Fellows conceived the gear shaper. Woolson backed Fellow's invention that led to the launching of the Fellows Gear Shaper Company in 1896. The plant, with Fellows as its first Manager, was established above the Black River Falls in Springfield, and maintains a thriving business today.

The first Fellows Gear Shaper, 1896. Credit Fellows Gear Shaper Co., Springfield, Vermont.

The Bryant Chucking Grinder.

William Leroy Bryant, who sought to eliminate friction both from machines and from human relationships, was by training and and instinct a machine-tool engineer. In the grinder department of the Jones and Lamson Company, Bryant found the testing ground for his invention that became the Bryant Chucking Grinder. To build this machine the Bryant Chucking Grinder Company was organized in 1909.

That year I was assigned to work on Bryant's project in the grinder department of the Jones and Lamson Shop. It was an area where I had had limited experience, and I asked to be spared such responsibility, but my request was not granted. One of Mr. Hartness's principles was to try a new man in a new department. Confronted with the situation, I contrived a system to help me learn the art of grinding. With my operators I studied every angle and detail, made a record of the facts and filed them for future use.

My good friend Frank Hill was succeeded as Shop Superintendent by Charlie Johnson[4] in 1909, two years after I began to work at Jones and Lamson's. Mr. Hartness held Charlie in high esteem.

Fred P. Lovejoy. Courtesy Mrs. F. P. Lovejoy.

Left, Charles W. Johnson, courtesy E. Johnson. Right, John E. Lovely, Chief Engineer and Vice President, Jones and Lamson Co.

He possessed a kind of negative asset that qualified him well for his new position. Not an inventor, he offered no professional competition to those who were. He did, however, possess a keen sense of what was mechanically appropriate. He served as Shop Superintendent until retirement in 1928.

One day while I was in charge of the Planer Department, Charlie stopped at my desk with a word of appreciation: "Oscar, I don't want to give you a swelled head, but you've turned out more work in less time and with fewer men than . . . ever did." And he was gone in a customary flash before I could react.

During the first decade of the twentieth century, Springfield's industrial strength expanded. Both Jones and Lamson and the Fellows Gear Shaper companies had enlarged their plants. Our shop was operating a night shift. By the end of that decade, however, we heard rumblings of the depression that threatened the country from 1911 until the outbreak of World War I in 1914.

Business grew slack. I made another attempt to write, but this effort failed for two reasons. The first was a bad case of the flu. The second was Mr. Johnson's announcement that I had been appointed to superintend a new area recently created in the shop. It was the Special Machine Tool Department.

The Management of Jones and Lamson Company organized the Special Machine Tool Department in 1912 under the direction of Fred Lovejoy.[5] Its purpose was designed to meet the growing competition created by mass production and automation. Both cut into Jones and Lamson territory. Fred Lovejoy, chief engineer of special tools, possessed marked ability for devising them. He was a man of integrity, discreet and deliberate in all his ways. That I had been his choice for superintendent of the new Special Machine Tool Department was not disclosed to me until long afterward. My most intimate associations during the years in that department were with this man of Lincolnesque stature, Fred Lovejoy.

Among the first-rate machinists with whom this department began were these: Walter Allen, Frank Whitney, Bill Brady, Fred Barber, John Bemis, and Everett Redfield. Some of these names will appear again as the story proceeds.

Our Special Machine Tool Department reserved space for a trial run of every outfit on a Turret Lathe kept in our department. Eventually, equipment for the Fay Automatic Lathe was designed with tooling for every type of job it was destined to handle.[6]

We were doing a good volume of business well into 1912, when a gray sunset presaged a long depression. It ended in 1914, not because of a wholesome revival in the industrial world, but rather to answer the demands of World War I.

The word that I was to head up the new Tool Department had come unexpectedly. Many machine-tool men would have welcomed this opportunity with eager acceptance. My own response fell short of enthusiasm in spite of its recognition of a certain fitness for the job. Where necessity dictates there is little choice. I assumed my responsibilities in April, 1914, but it was December before I felt a measure of energy to meet their demands. Never dreaming where it would take me, I had made my concession to the career of journeyman machinist.

Coming home from the day's work at Jones and Lamson.

Chapter VIII

SPADEWORK ON THE SIDE

I had come to terms with a career as journeyman machinist, but had not quieted the urge to write.[1] Through the years, while on the job in the shop, I observed the growing phenomenon of mass production and studied factors that contributed to its rapid progress. Following are some of my findings.

The development of steel, electrical energy, artificial abrasives, and the grinding machine helped to set the stage for industrialization in the twentieth century.

The origin of steel appears to be lost in antiquity. It is believed that "Damascus steel" may have originated somewhere between India and the east coast of the Mediterranean Sea, around 4,000 B.C. A sword of this ancient metal was reputed to cut gauze in midair and to sever an iron nail without a nick. Little wonder that Damocles paled at the sight of one of them hung by a horse hair above his head.

The Roman Emperor Diocletian fostered the making of the Damascus sword in the third century of the Christian era. The Mongol warrior Tamerlane carried Damascus smiths back to Samarkand of central Asia in the fourteenth century, at which time this sword of Damascus steel disappeared.

Five centuries later, the Russian General Anossof of Zlatoosk in the Urals created a sword that was supposed to have matched the performance of the Damascus steel blade. According to the Encyclopedia Americana (1940), Anossof made his sword from iron, iron oxide and graphite. Then he tempered the instrument in boiling oil. Perhaps this Russian inventor was a descendent of the Mongol invaders of Tamerlane's time. (Timur Lenk)

Sir Henry Bessemer, F.R.S. Courtesy Institute Mechanical Engineers of Great Britain.

Three men, contemporaries in the nineteenth century, made significant contributions to the manufacture of modern steel: Sir Henry Bessemer and Robert F. Mushet of England, and Koran F. Koransson, a German-born Swede.[2]

Bessemer discovered a process that reduced the production cost of steel. He mixed copious amounts of air with molten iron. When the iron cooled, the result was comparable to the one obtained by great quantities of fuel consumed to the detriment of England's forests. This was a revolutionary discovery.

Bessemer licensed the English steel-makers to use his method. Koransson of Sweden bought a license from Bessemer. Then failure hit. Steelmakers did not succeed with the Bessemer process and there followed a temporary setback.

Koransson began to experiment in Sweden. He used a force-pump to increase the volume of oxygen and met with success. He made thirty tons of this product. Half of it he shipped to Bessemer and the rest to other steel-makers in England.

Robert Mushet, a research student, anticipated the field of modern metallurgy. In his experiments with metals, he found that cast iron varied appreciably according to the ores from which it derived. He discovered that spiegeleisen, an ore with manganese, responded more readily than others to the steel-making process. He chanced upon an alloy combination, tungsten and carbon steel, for making high-speed cutting tools. In 1868, English patents protected his discovery. His tools found their way to the Continent and to the United States. He may well be called the grandfather of today's high-speed cutting tools.

Progress in the development of electrical energy gave further impetus to the advance of mass production. Large enterprises had successfully utilized it, but in smaller machine shops hysteresis developed. This leak, as it was called, baffled electrical engineers until Charles P. Steinmetz[3] solved the problem.

The manufacture of artificial abrasives, the artificial diamond in particular, contributed extensively to hasten mass production. It

Left, Koran F. Koransson. Courtesy Institute Mechanical Engineers of Great Britain. Right, Robert F. Mushet. Courtesy Institute Mechanical Engineers of Great Britain.

Left, Charles Proteus Steinmetz. Courtesy General Electric Co. Right, Dr. Edward G. Acheson, City Carborundum Co. Courtesy Francis D. Bowman, Carborundum Co.

is curious that three men were experimenting independently in the same field. Each one tried to create a hard substance that would cut glass and act as an abrasive as well.

Dr. E. G. Acheson, of Monongahela, Pennsylvania, experimented with sand, sawdust and coke. The result in 1891 was an artificial diamond. It went on the market to cut diamonds and polish gems.

Charles B. Jacobs achieved successful results with a clay known today as bauxite, which is found in Baux, France. Rich deposits of the material are also located in the State of Arkansas.

Aldus C. Higgins used still another method. He developed an electric furnace with temperatures high enough to fuse substances into abrasive material.

The gargantuan appetite of modern industrial processes would suffer were it not for the pioneering efforts of these men in search of the artificial abrasives created through their experiments.

A fourth development that helped set the stage for mass production was the invention of the grinding machine. With it is associated the name of Charles H. Norton (1851-1942). At the dawn of modern industrialization he anticipated mass production.

Norton had valuable experience in the machine-tool industry with the Brown and Sharpe Company of Providence, Rhode Island. He was called to the Norton Company in Worcester, Massachusetts. This firm, in no way connected with the Charles Norton of our story, expanded its plant to construct grinders.[4] In the process, Charles Norton's idea for building a specific, heavy-duty grinding machine became a reality. It proved to be an immediate success. Because of this man's engineering genius, the Norton Company set a pattern for making grinders that other shops were destined to follow.

The Norton Grinding Machine. Courtesy Norton Co., Worcester, Mass.

We have taken a brief look at four developments in nineteenth-century industry that prepared the way for mass production. Now let us consider the four wheels of this chariot (as I like to call modern industry) that carried mass production into the twentieth century: the grinder, gear-tooth generator, metallurgy and electrodynamics.

The first Norton Cylindrical Grinding Machine, 1900, presently in the Edison Museum. Courtesy Norton Co.

Charles H. Norton about 1916. Courtesy Norton Company.

The grinding process is as old as our earth. Nature used it with her glacier machine on our mountains and along the shorelines of our land. Primitive man discovered and used it in his everyday life. Modern man could hardly function without the developments that the grinding process makes possible. We have already mentioned the names of Charles H. Norton, and of William L. Bryant before him in chapter VII. Their inventions in the field of grinders helped to meet the demands of modern industrialization.

No gear of any description, so far as I can learn, is listed among archeological finds. It appears to be a modern device, a result of the cumulative effort of many minds. The water and paddle wheels were gears in principle. Eventually, man harnessed steam and electricity for power to drive his generators.

The names of three men are significant in the field of gear-generator machines: William B. Gleason, Edwin R. Fellows referred to in chapter VII, and William B. Sykes.

Gleason, born in Ireland in 1836, grew up in America. In 1874, he built his first bevel-gear machine, that is, a machine that transmits power around a curve. Later, the Gleason Works of Rochester, New York,[5] became renowned for their bevel-gear machines.

William E. Sykes, an Englishman,[6] received distinction in his manufacture of the herringbone or helical spur gear. It proved advantageous in heavy and high-speed machinery, notably turbine engines and oil-drilling machines.

 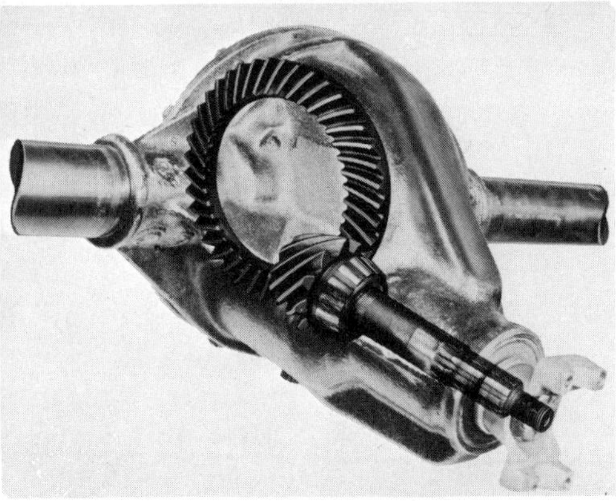

Left, William Gleason (1836-1922). Courtesy Gleason Works. Right, Hypoid Bevel Gear Drive, small cut. Courtesy Gleason Works, Rochester, New York.

Three problems confront the designer and the builder of gear-machines: strength and smoothness of operation, plus precise timing of the gears. To construct mechanical details from the engineer's most accurate drawing demands scrupulous attention. Variation in the timing gears is inadmissable for accurate performance of a machine. The precision of the wormgear drive of a telescope, for example, must be better than one-tenth of one thousandth of an inch for an instrument that is to follow a star millions of light years away.

Left, William E. Sykes of Farrel-Birmingham, Inc., makers of the herringbone gear, Ansonia, Connecticut. Courtesy Farrel Birmingham Co., Inc. Right, Farrel-Sykes Continuous Tooth Herringbone Pump Rotors. Courtesy Farrel Birmingham Co., Inc.

Metallurgy, third of our industrial chariot wheels, emerged as a theoretical subject from the professor's laboratory to enter, within a generation, the workshop of industry. We mentioned Robert Mushet's work in alloy combinations that anticipated the science of metallurgy. Modern industry would continue to move in nineteenth-century rhythm were it not for the machine tools of high-speed steel that metallurgy makes available for mass production.

With gathering speed, modern industry moved into mass production on the power of electro-dynamics, fourth of the chariot wheels. Noiseless and vibrationless, electrical power moves among the most sensitive mechanical devices from the dentist's instruments to the intricate machinery of Jacques Cousteau's underwater equipment, a spaceship, or the giant eye of the Hale Telescope on Mount Palomar, which explores the universe.

Our modern industrial chariot has a fifth wheel, an indispensable one that we call welding. This process grew from humble origins. The metal craftsman of the mid-nineteenth century worked in his little water-powered shop. He created and he repaired with forge and bellows, anvil and tongs, vise and hammer, chisel and file, hand-operated metal planer and blacksmith's drill-press.

On the heels of modern industrialization rode the demands for salvage and repair jobs. Nineteenth century methods of cutting and welding were no longer adequate. With progress in the science of metallurgy, oxyacetylene became available for use in salvage work and for the creation of new machines.

Interested in the origins of the oxyacetylene torch, I wrote a letter of inquiry to The Linde Air Products Company. In reply, I received a letter dated May 23, 1947, from Mr. C. T. Fallon of that Company, Unit of Union Carbide and Carbon Corporation, 416 West Eighth Street, Los Angeles, California. Following are some excerpts from that correspondence.

> The first reference in history to anything approaching modern gas welding was in 1820. Broke, a German, brazed two pieces of iron together in his laboratory with an oxy-hydrogen flame . . . In 1887-1888, an Englishman, Thomas Fletcher, presented several papers before various English Technical Societies, on the use of oxygen and hydrogen for burning out tap holes in steel furnaces. Among the first to take advantage of this process were safe-breakers . . .

> Le Chatelier, a French chemist, experimented with the oxyacetylene flame. From his experiments have developed the oxyacetylene welding and cutting . . . Everything necessary to oxyacetylene welding and cutting was available except the apparatus itself.
>
> This problem was solved in 1900-1901 by Fouche and Picard in France . . . They made the high pressure blowpipe or torch that utilized the two gases at equal pressure . . . Because of certain inherent drawbacks to the equal pressure system of mixing the two gases, Fouche attacked the problem . . . He finally hit upon the injector principle . . . In 1903, he had perfected his low pressure blowpipe and was marketing it in France . . . His right to the title Father of Oxyacetylene Apparatus has never been challenged . . . In 1904, Jotterand of Belgium succeeded in cutting heavy steel with an oxyacetylene blowpipe. He applied for a United States Patent on his process in 1905, and in 1906 the patent was issued on the cutting or severing of steel by the oxyacetylene process . . . Here was the most important tool ever invented for the steel industry and the patent authorities accepted it with alacrity. In less than five years it was mechanized for straight lines and circles. The cutting of intricate shapes followed rapidly, although it was many years before industry would accept this universal tool on a par with other machine tools.

One could write an epic about lives freed from imprisonment by the oxyacetelene torch.

With the quotation from Mr. Fallon's letter, I conclude my spadework into the roots of mass production. I turn now to the restless years around World War I and a leisure-time project that brought its rewards.

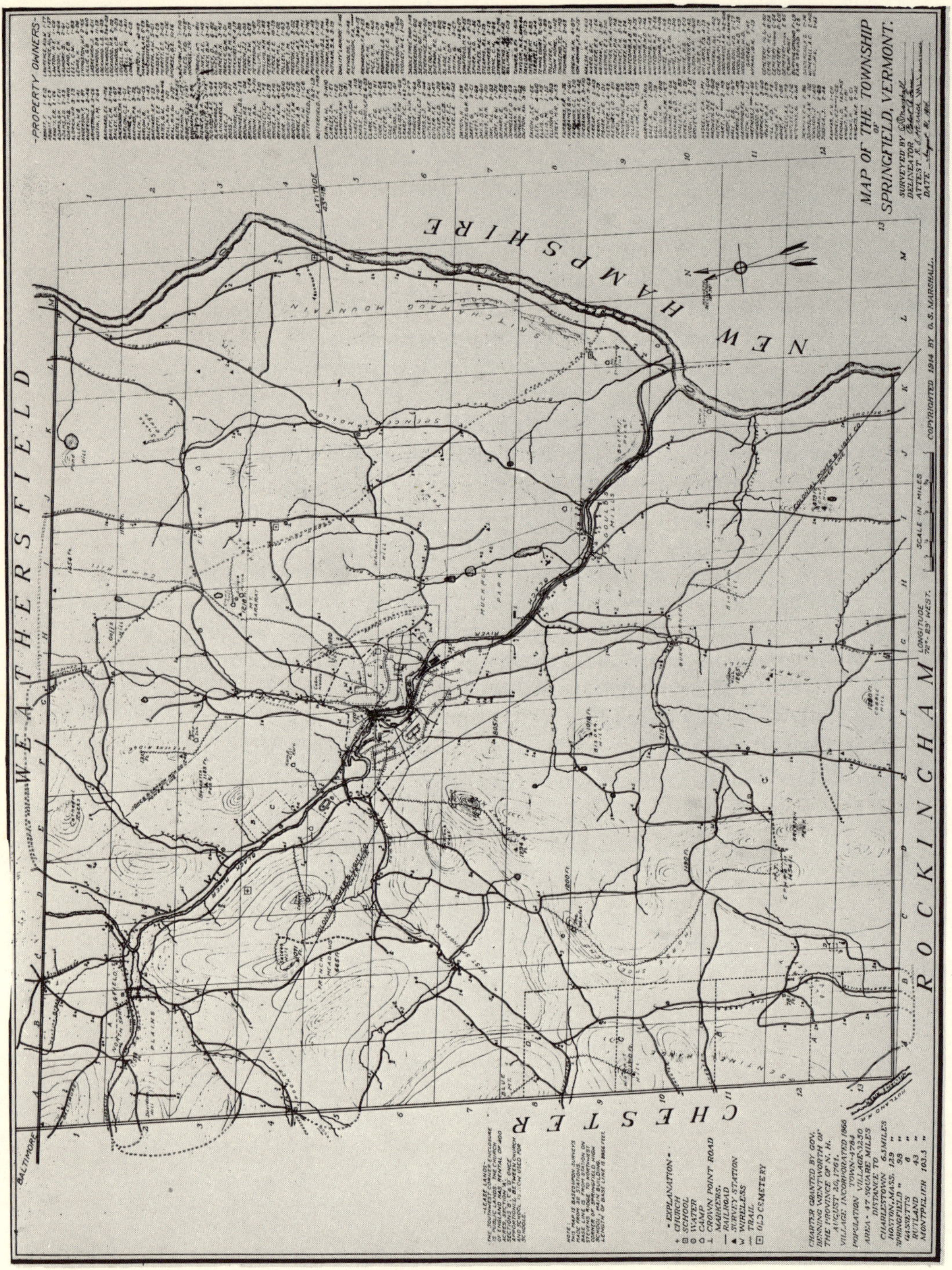

Map of the Township of Springfield, Vermont, surveyed by OSM in 1914.

Chapter IX

THE POTATO STONE

As I write these lines in 1945, the Potsdam Conference takes place following the defeat of Germany in World War II. One would turn his eyes away from human history, saying with a sigh, "Man is only a fighting animal. His mind is not awake to what life is all about."

My thoughts turn back to the days prior to World War I, when illness hit our family,[1] and a prolonged business depression reduced shop hours to part-time work. During those months of forced leisure time, in addition to writing a variety of articles,[2] I became interested in old Vermont maps. They inspired me to think about making one of Springfield. The government had not completed its geodetic survey of that area. This encouraged me to do so myself. Furthermore, I felt the call of the out-of-doors.

For the sum of $11.00 I had purchased an excellent forty-power refractor telescope,[3] sometimes called a spyglass. With my telescope plus an aneroid barometer and an altazimuth for surveying purposes, I carried a notebook and my lunchbag. Together we covered many miles on bicycle and on foot over the uneven terrain of the township of Springfield.

At the end of what totaled thirty days of eight hours each, in the field and at the plane table, I submitted the survey to Walker Brothers of Boston. They lithographed the map. Wheelers Pharmacy, the old corner drug store in Springfield's public square, assumed responsibility for its sale.

This map became the means of my introduction to many individuals. Among them was E. Wellman Barnard,[4] retired and living at his ancestral home on Open Ridge Farm, in the eastern outskirts of Springfield. He was an old war horse who, as correspon-

Oscar Marshall as an amateur surveyor in Springfield, Vermont, in 1914.

Oscar Marshall's forty-power refractor telescope used in surveying the Springfield Map, 1914.

dent for Boston newspapers, had toured Europe and interviewed several notables, one of whom was Count Zeppelin of dirigible fame. He and Barnard maintained a correspondence. As a result, plans developed for the Count to fly from Germany and land on the Open Ridge property.

Before the venture materialized, hostilities broke out in Europe. My boss, Mr. Hartness, at the Jones and Lamson Company, im-

Mr. Barnard's sugar house.

mediately advised that our Springfield map should be temporarily withdrawn from circulation. He had learned about the German Count's proposed flight-landing on the Springfield farm.

Kaiser Wilhelm's military aggression in 1914 drew us into the European conflict of World War I. I experienced the impact it made upon the Jones and Lamson Machine Company. The two-year business depression turned quickly into a period of productive activity in Springfield's machine-tool industries. Mechanics, who had taken up farming to relieve their economic situation, now swarmed back to the shops: Fellows Gear Shaper, Bryant Chucking Grinder and the Jones and Lamson Machine Companies. Their men oiled long-idle machines, reconditioned disabled ones and cleaned and rearranged their workbenches. "Lick the Kaiser and his Prussians" became the popular slogan.

The work force of Jones and Lamson Machine Company increased from 300 to approximately 700 during the peak of the war effort. The Special Tool Department doubled its number of men and its amount of equipment. I acquired both an assistant and a secretary. The workday was extended from eight hours to twelve for the regular operators and a large shift went on night duty.

Other developments occurred. The administration enlarged the shop with a new plant that included a restaurant with dinners served at cost for the working force. Jones and Lamson's took on the appearance of a city establishment.

One day, "W.D." as we called Mr. William D. Woolson, Treasurer of the Company, was out in the shop. With an inquiring expression on his face, he said that he was looking for a certain department. He wasn't familiar with the new machine-shop arrangement. When I explained just what had recently taken place, he told me this story.

"Pete and Pierre were night-fishing on Lake Champlain. They anchored and both fell asleep. Several hours later, Pierre awoke, looked about, and noticed that their anchor hadn't held.

He called out, "Pete, Pete?"

"What is it?"

"We ain't here!"

Pete looked around a minute and remarked "By gar . . . we ain't here, be we, we're two mile from here!"

About this time the young managers of the Company felt that our local weekly newspaper should include what they called a *Jones and Lamson Column.* They asked the editor to solicit me for material. I had to decline because of my responsibilities in the Special Tool Department.

Mr. Hartness, however, heard about my refusal to write for the paper. He did not accept that as final. I had written an article entitled "An Interpretation of the Human Factor in Works Management" to be read at a foremen's meeting. To make certain of his approval, I showed it to Mr. Hartness. He was pleased with the paper and added that I should pursue my efforts in the field of writing.

With regard to the shop column in the local paper, Mr. Hartness was not to be put off. He accosted me with the statement, "We must have that shop column in the weekly paper and you must take charge of it."

"If it means helping with the war effort, I will undertake it, but not for any other motive," I replied.

The dual load proved too heavy. We made arrangements for my shop assistant to assume temporary supervision of the Special Tool Department.

As Editor of the Shop Column in The Springfield Reporter, I found a real measure of satisfaction.[5] Also as Editor, I proposed a celebration for Charlie Johnson's sixtieth birthday. Mr. Hartness was enthusiastic about the plans. Both Management and Shop Foremen were to participate. Hartness stipulated that Frank Hill,

Shop Superintendent from 1904-1909, and whose abilities he had recognized, should be invited. Ten years previously a rupture had occurred between Mr. Hartness and Mr. Hill, over a minor matter, but in the end Mr. Hill was discharged and Charlie Johnson succeeded him as Shop Superintendent.

The birthday celebration testified to the repentance of Mr. Hartness and the forgiving spirit of Mr. Hill. Mr. Hartness presented a fine gold watch to Charlie Johnson, who acknowledged the gift with appropriate words. Then Mr. Hartness asked Frank Hill to stand. After a few words, Hartness offered to him, who had been his fourth Shop Superintendent, a fine gold watch also. Whereupon Hill astonished the party with this story.

"A widow was visiting the grave of her fourth husband. As she wept bitter tears, she caught the sound of a voice beneath the sod and this is what she heard, 'Well, if I'd known you thought so much of me — damned if I'd a died.'"

After World War I, the machine-tool business went into the doldrums. Even so, no bread lines developed in Springfield. Her citizens, machinists and mechanics included, turned again to Mother Nature and her natural assets in soil and livestock.

With the post-war slump that affected our shop life came a marked change in the general pattern of the machine-tool industry. It appeared in the altered role of the journeyman machinist, who was being replaced by the lathe man, milling man, planer, gear-cutter, bench hand and so on. The workman whose function most closely approached that of the journeyman machinist was the instrument-maker. He wore a long white coat, necktie and polished shoes. His spruced-up appearance set him apart from the run-of-the-mill workmen. It was easy for a mechanic to feel more than a tincture of inferiority complex at the sight of his long white coat.

Automation was affecting industry in that the war effort had already accelerated mass production. Engineers[6] were hard-pressed by new demands for the manufacture of automobiles and airplanes. At Jones and Lamson's, Ralph Flanders had invented a thread grinder, the precursor of the present unique machine. I had been given charge of building the first ten of the new machines. Ralph's brother, Ernest, supervised the production department. In spite of certain setbacks, the Universal Thread Grinder of the Flanders brothers met a flattering success and is without a challenging competitor in its field.

The Edward Longstreth Medal awarded by the Franklin Institute of Philadelphia April 15, 1942, to the brothers Ralph E. Flanders, left, and Ernest V. Flanders, right, for the development of a thread grinding machine that included a method for truing the grinding-wheel.

This restless journeyman machinist, however, faced those uncertain twenties with something less than enthusiasm, until the day when he met Russell Porter.

In 1919, Mr. Hartness brought Mr. Porter to the staff of Jones and Lamson's Machine Shop, where he made a significant contribution to the Hartness Comparator, an optical inspection instrument. He had come from the Bureau of Standards' Optical Branch in Washington, D. C.

Upon Porter's arrival, Mr. Hartness gave him a tourist copy of the Springfield map. One day, Robert Beardslee, also an associate on the comparator job, conducted him to my desk and introduced us. Porter held the map in his hand and began at once to question me. He was curious to know how I had determined the base line and asked if I would consent to his checking it with me. I willingly agreed.

I should say here that a detailed description of the measuring process used in making the Springfield map appeared in *Popular Astronomy,*[7] March, 1922, under the title, "The Telescope as a Measuring Instrument."

After shop hours, we went together while Porter re-checked my findings. He made use of his theodolite, a precision instrument that he had used when surveying in Alaska with Dr. Frederick A. Cook in 1906.

Within a week, Porter came to tell me that he found a difference of less than one foot in a thousand between his results and the base line as printed on my map. Then he said, "If anyone questions that base line, send him to me."

Russell Porter's arrival at Jones and Lamson's in 1919 marked the birth of a new enthusiasm for me. It developed in the exciting

James Hartness and Russell W. Porter about 1920.

journey with this man Porter who led a group of us in Springfield through the intricacies of telescope-making to the rewards of star-gazing. It also found nourishment during that brief interlude in the early twenties when Jones and Lamson produced The Porter Garden Telescope.

A by-product of what came to be known as The Poor Man's Telescope was The Porter Garden Telescope. It was built originally to gratify the Springfield artist, Horace Brown, who had expressed the wish to have an ornamental telescope.

This Garden Telescope was a departure from the normal product made by Jones and Lamson's, but both Mr. Hartness and Mr. Flanders were in favor of the enterprise. They felt confident that the ornamental instrument would appeal to wealthy owners of fine estates. Hartness and Flanders were right with respect to the appeal of this decorative telescope, but they misjudged the ability of the owners to use it competently for its designed purpose.

With the rare skill for which he became known, Russell Porter designed the instrument that was patented September 25, 1923. I was put in charge of building the mechanical structure, an assignment that gave welcome relief from the boredom of my usual shop days. Russell superintended the optical works. Plans were laid to market this new member of the telescope family. The outlook was promising. Orders arrived so promptly that we had to step up the tempo of production.

Gradually the need for a field agent became apparent. I was delegated to visit each party who had purchased a Garden Telescope. My assignment was to help him achieve 100 percent satisfaction with his new instrument. In every instance I discovered that the owner faced identical difficulties and for the same two reasons. First, he had failed to heed the printed instructions concerning the care of the mirror. These had been carefully enclosed with the instrument. Second, the operation of the reflecting telescope was complicated at best, in spite of detailed information which Russell had provided. The instrument, designed for astronomical use, failed to be effective in the hands of those who were unskilled in basic astronomy. When the Porter Garden Telescope was discontinued after two years, one of my shop friends remarked, "Oscar has had to go back to work."

The title of this chapter may need an explanation. Potato stone is the nickname for a geode. Shaped like a small potato, it may be half as large as a football and is found on the surface of the ground or not far below. The geode's rigid rock formation conceals a cavity lined with beautiful crystals.

Like the geode, my achievement of the Springfield Map concealed gems of experience that came to light a dozen years later.

Oscar S. Marshall on a mountain climbing day.

The Old Observatory at Russell W. Porter's seacost home Land's End, Port Clyde, Maine. Pencil drawing by OSM, August 2, 1922.

Chapter X

HOPE ON THE HORIZON

His full name was Russell William Porter, but to the world of amateur telescope makers he was Russell, as he was to his friends and associates. He is worshiped by thousands of amateur makers of the "poor man's telescope." The word worshiped is not too strong. It expresses the attitude of that great band of persons whom he enabled to look into the heavens with their own home-made instruments.

Others[1] have written about him, but this is a personal account of the man as I knew him. He was born December 13, 1871, in Springfield, Vermont. Russell, cast in his individual mold, was not a harum-scarum youngster, but his father kept a careful eye on him, as the following incidents will illustrate.

Russell found a tempting challenge in the undergirding of the new steel bridge that spanned the Black River Falls roaring through Springfield. He delighted in trapeze performances on the lower structural framework of that bridge until his father discovered his antics.

He also tried his aim at hitting one of the small windowpanes of the barn gable. Upon detection, Russell received a handful of pebbles from his father who led him to a point between the barn and the house. Standing a few feet away, the elder Porter instructed his son to see how close he could come to hitting him. "That just about broke my heart," Russell confessed to me.

Russell graduated from Springfield High School. He explored the possibilities of the University of Vermont and of Norwich University before he found what caught his imagination at the Massachusetts Institute of Technology. There he enrolled to study art and architecture. He won first prize for an original architectural design that he submitted in a nation-wide competition sponsored by the Society of Beaux Arts Architects.

The Arctic regions held immense fascination for Russell. Although not a member of Peary's Arctic expeditions, he did participate in one to Greenland in 1896. His magnum bonum came when, as artist and assistant scientist, he was second in command to Captain Anthony Fiala of the Ziegler North Polar Expedition 1903-1906. His diary[2] of that trip is an impressive account. The following incident was recorded during that time.

Because of the severe Arctic weather conditions, the relief ship could not get through to Captain Fiala's party and they had to spend a second winter on Franz Josef Land. The experience brought out the stamina of every member of the expedition.

On one occasion, when there was an extended period of daylight, Russell decided to make some sketches a mile or so from camp. He took his water colors and a small alcohol lamp to keep the water soluble. He carried along his 45/70 Marlin repeater in case fresh meat should appear from somewhere.

Absorbed in his painting, he happened to look up and saw a polar bear but a few rods away. Russell maintained a rigid silence, nor did he move, hoping that the great beast would alter his course. He didn't. He moved slowly swinging his head as if to locate the source and nature of some strange odor. Increasingly curious, he followed his nose in the direction of the alcohol lamp.

Russell concluded that here was fresh meat for the camp. He aimed his Marlin at the advancing mass of white, but was unable to work the breach mechanism to force the shell into its chamber. The bear ambled steadily closer. Then Russell got the stubborn shell 'home' and fired at the bear. The creature continued toward his objective.

Reloading the Marlin was all but impossible. The infuriated bear was marching in for the kill. This was no time to lose one's head. Russell worked at the lever, banged the gun against his knee and the shell went into its chamber. He fired his second shot and the animal dropped at Russell's feet.

The bear's upper jaw remained Russell's only tangible souvenir of that particular encounter. If the polar monarch instead of the artist with his Marlin had been the victor, the rest of our narrative, indeed the most significant part, would never have been written.

In addition to his survival of a variety of crises encountered in the frozen northland, Russell experienced a rebirth there, where celestial phenomena became for him the object of an all-consuming interest. As assistant scientist on the trip with Captain Fiala, he had used a transit instrument to record time during the vigil of the long Arctic night. He resolved that some day he would make his own telescope.

You might think that Russell's appetite for Arctic life would have been satisfied. Not so. In 1906, he joined Dr. Frederick A. Cook, American explorer, in his expedition to conquer Mt. McKinley. A report of Russell's topographic survey, made at that time, is in the Yentna District Alaska U. S. Bulletin 532 for the year 1913.

With these Arctic exploits behind him, Russell began to dedicate much of his time to astronomy. Since he could not afford to buy a good refractor, he worked on plans to make his own instrument, a reflecting telescope. He set about this new and absorbing subject as he always did a worthwhile problem — methodically and thoroughly. "With my two hands and my brain I will undertake this task," was the motto that he occasionally uttered for another's encouragement.

Before 1919, when I first met Russell at Jones and Lamson's, he had married Alice Bell Marshall of Port Clyde, Maine. Their seacoast home, known as Land's End, became a retreat for them and a welcome rendezvous for their friends with kindred interests. I was fortunate to be their guest on several occasions when I saw relics reminiscent of Russell's efforts to build a reflecting telescope.

Among those reminders of his early struggles were (1) his canvas-covered wooden observatory, (2) his indoor observatory created by cutting a hole through the southeast corner of his in-law's house plus a chimney-like adaptation for his two mirrors (These are the major optical units of the Porter Indoor Observatory at Stellafane, Springfield, Vermont), and (3) a 175-foot passage, his hand-hewn tunnel through dense cat-spruce foliage close to the ground. It offered the near-total darkness required to give to his mirrors the Foucault knife-edge test. This would reveal their parabolic fitness for the reflecting telescope.

Russell belongs among those distinguished amateur builders of the reflecting telescope, self-made experts such as the professional musician, Sir William Herschel, the painter, Alvan Clark, and the steel-rolling mechanic, John Brashear.

A new glow brightened my horizen in 1919 when Russell Porter returned to Springfield.

Charter members of the Springfield Telescope Makers, Springfield, Vermont. Left to right: OSM, Clyde P. Baldwin, F. Eugene Lockwood, Russell W. Porter, Raymond P. W. Fairbanks, Gladys Piper, Oscar P. Fullam, Ralph Baker, Frank H. Whitney, Roy J. Lyon, Everett H. Redfield, Carlton B. Damon, Ernest N. Brookings, Guy E. Baker. Absent from picture: Albert A. Herrick, Charles A. Longe, John M. Pierce. October, 1920.

CHAPTER XI

EPIC OF THE NIGHT

Not long after Russell came to work at Jones and Lamson's Machine Shop, he and I began our frequent outings together. We camped as the mood hit us. We found each other cogenial company.

Within the year, Russell discovered that some of the shopmen showed interest in astronomy and in making reflecting telescopes. Imbued with a kind of evangelistic spirit, he wanted to organize a class made up of Jones and Lamson mechanics and show them how to make mirrors for their own instruments and how to build their equatorial mounts. The only compensation he considered acceptable was the full cooperation of each individual.

It became natural for me to act as Russell's liaison man in this new development. I enrolled sixteen persons for our first meeting August 17, 1920. Among these were one woman, Mrs. Gladys Piper, and one professor, John M. Pierce, head of the Springfield High School Cooperative Department. The rest were mechanics, engineers and foremen of Jones and Lamson's.

These persons formed their organization under the name of The Springfield Telescope Makers. Charter members were: Russell W. Porter, Founder and Director; Oscar S. Marshall, Secretary-Treasurer; John M. Pierce, Oscar P. Fullam, Everett H. Redfield, Frank H. Whitney, Ernest N. Brookings, Raymond P. Fairbanks, Roy L. Lyon, Albert H. Herrick, Mrs. Gladys M. Piper, Fred W. Barber, Guy E. Baker, F. Eugene Lockwood, Clyde P. Baldwin, Charles N. Longe, and Carlton P. Damon.

Five of us were over fifty: Redfield, who had made two Grandfather Clocks; Whitney, who was an exceptional mechanic; Fullam, an expert pattern maker, who proved his skill with designs for our equatorial mounts; Russell, and myself.

Russell ordered our materials: two discs of glass for each member, carborundum of several grades from coarse to fine for grinding purposes, rouge for polishing and Wilmington pitch for laps when we used rouge.[1]

Our mirrors varied from six to nine inches in diameter. They were correspondingly thick to maintain freedom from distortion.

Russell set us agog at the outset by saying that while accustomed to working to a few tenths of a thousandth of an inch, we were now embarking on the conquest of dimensions in the realm of millionths of an inch. This amazing refinement would confront us when we applied the Foucault knife-edge test to our mirrors. No scattering of shots at the target, as the saying goes, for a well-defined image.

The reflecting telescope requires a single mirror that may be made of cheap plate glass, a factor in its favor. The refractor telescope mirrors must be made of selected optical glass.

By the summer of 1921, most of our group members had completed the grinding and polishing processes of their mirrors. Not all of them had finished their mounts for the instruments.

In August of that summer, we held our first exhibit at the local Town Fair. The reflecting telescope, an unfamiliar mechanical contrivance, aroused curious interest. Visitors were puzzled. "How can an object in the landscape be seen by looking anywhere but right at it?" they asked. Fullam's mount, made of seasoned maple, received critical study. Viewers concluded that the machine, as they called it, was either a churn or a cream separator.

Russell possessed an inexhaustible store of ideas that he was ready to share. With our group of eager amateur telescope makers, he found the right situation to do so, and he set about making plans.

First, he made certain that we were securely welded together as a group. To this end we staged all-night encampments on a hill in the vicinity. A bean-hole feast with plenty of baked beans (specialty of Everett Redfield, bachelor cook and poet) and roast corn, ham or bacon and coffee comprised the evening meal. For breakfast, there were substantial servings of bacon and eggs with toast and coffee.

Second, Russell recognized the need for organizational headquarters. Plans unfolded and a site was selected on Breezy Hill. He revealed his financial ability in securing funds for the clubhouse that we built together and eventually called *Stellafane*.[2]

The observatory at Stellafane under construction, with Mt. Ascutney and the Green Mountains in the background.

Russell W. Porter's sixteen-inch Indoor Reflector at Stellafane.

Cover of "Antares" reflector in Marshall backyard, winter 1929.

Our enterprise progressed during that period of slack business after World War I.[3] We found time to work on our instruments and to attend to matters that gave permanence to our new organization.

I, meanwhile, continued to work on my eight-inch reflector and its mount. Oscar Fullam, a genial individual and gifted, assisted me in building the cement pedestal, which was hexagonal in form and about five and one-half feet high.

This telescope, which I called *Antares*,[4] became the precursor of the so-called Springfield Mount that Russell perfected. Many amateur telescope makers throughout the country have used it.

The Springfield Mount makes use of a second prism placed outside the tube. It picks up the image from the main prism within the tube. The second prism is associated with the eyepiece that rotates for convenience. The observer may remain in a fixed position.

By fall of 1923, *Antares*[5] was in place on the pedestal. Its accurate orientation allowed us to pick up stars of first magnitude in broad daylight. From the ephemeris I found their locations and set the instrument with its right ascension and declination circles.

Oscar Fullam with his reflector.

Oscar Marshall with his "Antares".

"Antares" reflecting telescope, 1926.

Late in the summer of 1924, Mars was 34,000,000 miles from Earth. More than 400 visitors came to our home at 135 Wall Street to view this planet and the moon as seen through our eight-inch reflector. On September 4, Russell made further observations and drew sketches of Martian details, preserved in the Record Book at Stellafane.

After the completion of our clubhouse, we began to consider the possibility of a convention to receive amateur telescope makers from out of town. There were scores of them, especially in the eastern part of the country and Canada. Albert G. Ingalls of *The Scientific American* played a significant role in this growth of the Poor Man's Telescope Movement.[6]

The first convention took place in July, 1926, at Stellafane. The event became an annual affair in the month of August and grew to unexpected numerical proportions for two main reasons. One derived from the charm of an outing offered by the site of Stellafane itself, however humble the accomodations. Second, the visitors were eager to see the Founder of the Poor Man's Telescope, Russell W. Porter. They wanted to meet him face to face.

By the time of the third convention in August 1928, the amateur movement had reached wide geographical proportions. Some conventioners arrived from as far away as Kansas and Panama, while others came but a short distance. A comprehensive cross section of persons was represented by highway department workers, various types of white-collar employees, and professional people, among whom was the United States Navy Engineer W. E. Cloyd. He presented the parent organization with a gift of $100.00 to be used for its growing library.

We were amazed to see the ingenious creations brought by these visiting amateur telescope makers. For example, one of them used a simple spruce board six or eight inches wide, an inch thick and five or six feet long. To this he attached the optical parts of his instrument supported on a swivelling yoke. It reminded one of the old well sweep with its bucket carried on a long pole. With this contraption, our visiting amateur could look at the stars. The amateur mounts ranged from this crude construction to a fine equatorial mount of rare workmanship.

To some of these amateurs, thrills came because of what they had seen in the night skies with the aid of their own handiwork. To others, the thrill derived from the actual building of a successful instrument. Gathered at one of Nature's loveliest creations, these amateur telescope enthusiasts found their pent up desires released at Stellafane, when Night began to dim the earth and her stars slowly appeared. Most of the visitors came equipped to set up camp. Each one chose a site among the spruce trees surrounding the clubhouse, but few slept that one night of the convention. There was too much to explore and too much to be shared.

During that Epic of the Night with its music of the spheres, none of us dreamed that within less than a dozen years all of us would be involved with martial airs. The annual conventions at Stellafane continued with a growing participation until the summer of 1939, at the outbreak of World War II.

After convention dinner at Stellafane, 1937. At head of table is R. W. Porter, to his left Oscar Marshall, and then Ernest Flanders. John Pierce is to the right of Porter.

The Government then faced an emergency to which an army of amateur telescope makers responded. The War Department discovered a serious lack of roof prisms made from optical glass. They were part of the sighting apparatus of large guns, cannons, etc. The demand for them was imperative but no facilities existed, public or private, to meet that urgency.

Russell Porter, with Mr. Albert Ingalls of *The Scientific American* and one amateur telescope maker, Alex MacTavish, contacted the War Department officials. This trio offered to enlist the services of as many amateur telescope makers as practicable to make roof prisms.

The War Department officials were not sanguine about this source as the last word in optical refinement. Nevertheless, they agreed to give the offer a try. Failure would leave the situation no worse.

To the amazement of those officials, positive results were evident within a very short period, when roof prisms began to arrive from the four points of the compass. They came, not from professional workers in optics, but from amateur telescope makers, master technicians with their own techniques.

Mr. Albert Ingalls had been closely associated with the Amateur Telescope Makers Organization since 1925. From him,[7] I learned that, at the time of the War Department's emergency, there were over sixty amateur telescope organizations in America alone, not to mention a number of less well-defined ones in Europe.

That army of Amateur Telescope Makers, who volunteered to help meet the critical need for roof prisms, included the following: an accountant, a biologist, a cabinet worker, a candy maker, two chemists, a dentist, an economist, several engineers, a geologist, a gravestone maker, a metallurgist, three physicians, two physicists, and a steel worker.

The volunteer group of one single community was comprised of a variety of professionals: an herpetologist, a microscopist, a paleontologist, a telephone engineer, and a teletype engineer.

The annual convention of the Amateur Telescope Makers continues to take place at Stellafane, where the Epic of the Night grows longer every year in the lengthening shadow of its founder Russell W. Porter.

Chapter XII

A NIGHT ON MOUNT ASCUTNEY

Early in October 1928,[1] Russell and I arranged to spend the night camping on Mount Ascutney, which is situated in the towns of Weathersfield and Windsor, Vermont. Although this mass of granite offers no particular interest to the astronomer, geologists have puzzled over its formation.[2]

As an amateur photographer, I sought on countless occasions to capture its majestic beauty at different hours and from various places.

Mount Ascutney with Connecticut River in foreground near Windsor, Vermont. Photo by OSM.

As a lover of mountains, I had climbed the arduous trail of this monarch of the Green Mountains many times, from its base at 350 feet above sea level to its truncated summit at 3,300 feet. The exhilaration offered by the view and the inspiration of the quiet expanse were immense rewards for the struggle upward.

It was toward this lofty retreat that Russell and I set out in his Model-T Ford that October afternoon. At Brownsville, we left the car and began the rugged climb. After two hours we decided to bivouac on the shoulder of Mt. Ascutney for the rest of the night, but we didn't sleep a wink. Guess why. Russell's mind was busy with thoughts about a big telescope. He found a large piece of white birch bark. From the fire he salvaged a twig with a charred end to use for sketching. We needed illumination, but had left the flashlight in the car. I was lucky to find an old pine branch with a residue of pitch that burned. I held the torch for Russell while he drew his design of a huge telescope on the sheet of birch bark.

Before dawn and after a cup of coffee, we felt our way up the last steep rise to the top of Ascutney in time to see the sun

Caricature sketch by Paul Marshall Douglas. Oscar Marshall and Russell Porter on Mt. Ascutney, sketching their imaginary telescope on a piece of birch bark, October, 1928.

break over the New Hampshire Mountains. The air was frosty, the skies immaculate, and the horizon clear in every direction. Renewed by these moments of inspiration, we retraced our steps to the car at the base of the mountain for a real breakfast of eggs with bacon, more coffee and sandwiches. Then we headed back to Springfield.

Upon arriving in town, Russell discovered that Mr. Hartness, formerly Governor of Vermont,[3] had been trying to reach him. He was to report at once to the Hartness home. He arrived there without having changed his apparel of the last twenty-four hours, steeped in the odor of his stogie stub and the aroma of our mountain campfire.

Governor Hartness met Russell at the ring of the doorbell, saying, "Two gentlemen from California have been nervously waiting for over an hour to see you."

The Governor lost no time conducting Russell to the room where the visitors were seated, and introduced him to these men of science: Dr. John A. Anderson, Executive Officer of the proposed 200-inch telescope and Dr. Francis G. Pease, also associated with that project.

After the introductions, Dr. Pease spread a roll of large-scale blueprints on the table. They represented the proposed design for an enormous reflecting telescope. The four men went over those drawings in considerable detail, with the conclusion that they presented a workable plan.[4]

I learned later from Russell that, at the Bellows Falls Railroad Station where he and Hartness had accompanied the visitors, Dr. Anderson had asked him this question, "In case we want you to join the designing staff of our 200-inch telescope project, at the California Institute of Technology in Pasadena, would you consider the prospect?" Russell, who had caught the green light in the Governor's eye, had replied that he would be happy to do so.

On their ride back to Springfield, these friends discussed the reason that might have sparked the interest of the Cal Tech men in the services of Russell. Had it derived from his article[5] about the suggestive horse-shoe ring for the polar-bearing, which had appeared in the March 1918 issue of *Popular Astronomy*? Files of this periodical were preserved at the Mount Wilson Library in Pasadena, California.

Dr. Anderson informed me, while I was preparing this chapter, that Dr. George Ellery Hale, Director of the 200-inch Telescope Project, had heard about Russell's talent and his interest in designing telescopes. Dr. Hale aimed to explore every known source that pertained to the realization of his great dream and thus had instructed Anderson and Pease to contact Russell Porter in Springfield, Vermont.

A few weeks later, Russell received word from Dr. Hale that the Observatory Council would like him to join its staff of designers for the proposed 200-inch telescope. It was arranged, for the time being, that he devote six months each year to the telescope project and six months to his work at Jones and Lamson's where he retained his title, Associate in Optics and Instrument Design.

The Observatory Council members were informed that they were acquiring in their new associate a man with skills in architecture, art, mechanical engineering and surveying. Each talent was an essential asset for the great undertaking at hand. Furthermore, they were embodied in a person whose character would be a coordinating spirit, one who sought to create and to perpetuate harmony wherever his hand and his brain were in any degree responsible.

Thus it happened that, in so short a time after our night together on Mt. Ascutney, Russell attended a farewell reception in his honor. The entire membership of the S.T.M. (Springfield Telescope Makers) gathered at Stellafane, November 24, 1928, to celebrate the triumph of their Founder, Director and first President. Having had an indirect role in making this possible, we all shared the feeling of high achievement with which he accepted his new call. When he resigned as our president of the Springfield Telescope Makers, we unanimously elected John M. Pierce[6] to succeed him.

A few days later, Russell, with his wife Alice and their daughter Caroline, left by train for Pasadena, California.

CHAPTER XIII

ESCAPE

The Springfield Telescope Makers' organization was more than eight years old when Russell went to his new assignment in California, in November, 1928. We missed him in our midst. He had succeeded as he planned, however, in welding us into a body of self-reliant members. We met regularly and with our instruments we studied the heavenly bodies in their courses. We also kept in touch with other amateur groups dedicated to this mutual hobby.

Because of my restlessness with the daily grind on the job at the machine shop, I managed to find time to write,[1] and to enjoy photography, another hobby that I had pursued for a number of years.

Meanwhile, with the growth of the amateur telescope movement born in the area, a new interest in astronomy began to grow in the local community. It occurred to me, in the winter of 1930, that I might serve as an intermediary between the amateur astronomer and the layman curious about telescopes and the night skies.

As a modest beginning, I put a notice in the *Springfield Reporter*. It was an invitation for a limited number of young people to meet at our home one evening a week to consider the subject of astronomy.

Shortly thereafter, I received a most unexpected letter from Mrs. James Hartness, wife of the former Vermont governor. She made two suggestions. In the first one, she offered the use of her parlors as a place for the meetings proposed in my newspaper notice. Then she suggested that the invitation be extended to include a larger number of persons than I had indicated.

My original intent had not been that ambitious. Not acquainted with Mrs. Hartness, I regretted having exposed myself to her solicitous attention. I consulted Ernest Flanders,[2] a close friend, and one of the younger members of the Springfield Telescope Makers organization. He assured me that I should not hesitate to welcome her interest and cooperation.

Before making further decisions, I called on Mrs. Hartness and together we made plans. As a result, the following week a different notice appeared in the local newspaper. It announced that open meetings on astronomy would be held once a week at the Hartness residence.

I enlarged upon my earlier objectives in order to meet a wider range of interests and a larger number of participants. I kept three goals in mind: (1) to dispel feelings of inferiority with confidence built on solid information; (2) to implement facts with visual aids; and (3) to make use of simple language divested of technical phrases insofar as possible. With these goals in mind I worked out programs for a series of evening sessions.[3]

Four individuals contributed to the enrichment of these laymen's meetings. A member of the Springfield Telescope Makers organization gave a demonstration on the spectograph. Mr. Albert Ingalls, of *The Scientific American*, told us of his visit to prehistoric caves in France. The Reverend Augustine Jones, pastor of the local Congregational Church, shared some of his thoughts about man's relation to the cosmos and his destiny in it.

Undoubtedly the most popular event in the winter series took place on the evening when we were special guests of our fourth speaker, Mr. James Hartness, a long-time student of astronomy. He invited the members of our evening class to visit his indoor observatory, where he had built, in 1911, a turret telescope, a refractor with a 10-inch lens.[4] It was a special privilege for his guests to look at the planet Jupiter through that unique instrument.

In this personal venture, which provided me a welcome escape from the tedium of the machine shop, I wish to acknowledge appreciation to Mr. and Mrs. James Hartness. Alert for occasions to serve their community, they gave me the rare opportunity to do likewise in a way that appealed to my modest abilities.

Chapter XIV

THE CHALLENGE

In mid-spring of 1930, when preoccupied with the evening classes in astronomy, I received a letter from Russell in California. He asked if I would like to be one of the mechanics on the 200-inch telescope project at the California Institute of Technology in Pasadena. He added that, if I liked the work and if in turn the shop superintendent liked me, this might become a permanent situation. Without undue delay, I sent him an affirmative answer.

We kept this unexpected and exciting possibility a family secret until the plan with its details was definite. When Russell came back to Springfield for a brief visit in July, Mabelle and I made our revolutionary decision to head for new horizons, in September 1930.

Each member of the family felt the challenge as we all prepared for adventures that lay ahead. They began that September when, one by one, we left the Springfield home in which Mabelle and I had lived for more than twenty-five years. All of us did not leave at the same time, nor go in the same direction. Children usually are the first to fly the coop, but in our case the parents were the first to leave the nest.

On September 15, 1930, with countless details behind us, Mabelle and I, with my brother Eaton, set out in a three-seater Studebaker for Pasadena, California. To leave when we did, before the girls were ready to go, was not an enviable experience.[1] Beulah planned to travel north for her second year at the University of Vermont in Burlington; Candace would go south to begin her teaching career in New Jersey; Eva was scheduled to sail east in November for her life work as a missionary in North Africa. And we set out for the west coast.

Oscar and Mabelle Marshall in the summer of 1930.

Oscar and Mabelle with daughters Candace, Beulah and Eva, summer, 1930.

Eaton Marshall in the Petrified Forest, Arizona, September, 1930.

Grand Canyon, Arizona, September 1930.

In the year 1930, we spent fifteen days on our cross-country journey. Not until we had crossed the Mississippi River and the States of Missouri and Kansas did we begin to sense the enormous extent of our country. It was a challenge to the traveler's perseverance in the early nineteen-thirties.

At Trinidad, Colorado, where the Rocky Mountains became dimly visible, we began to feel that we had made some progress. Turning south to cross Raton and McAvoy passes, we descended through wilderness country to Taos, New Mexico, and spent the night.

By dawn the next morning, we were on our way again. I was at the wheel when we saw a large white sign in the distance. Gradually we made out the words: SHARP TURN LEFT. Sure enough it was both a sharp turn and a left one with a breath-taking curve. The highway had been carved out of the precipitous walls of the Rio Grande Canyon. No protective devices stood guard between us and the sheer drop to the tumbling Rio Grande River 1,000 feet below. I smothered an insane impulse to leap; but fully awake by this time, I managed to skirt the brink of the canyon.

The few hours spent in Santa Fe, New Mexico, rewarded us for not having by-passed that historic city. In Albuquerque, we took time to reload our film packs for the camera. Then we pushed on to visit the Petrified Forest and the Meteor Crater. In Flagstaff, Arizona, Dr. C. P. Lampland received us at the Lowell Observatory,[2] established by Percival Lowell in 1894. It is a center for special planetary studies, particularly of Mars.

Mid-afternoon Sunday, we arrived at the Grand Canyon. Only a daring spirit will essay with pen or brush to express for others his own reaction upon seeing that awesome wonder, a symphony of form and color in whose presence silence alone is appropriate.

One night remained before we finished the final lap of our journey to the west coast. We stayed at a camp site on Baldy Mesa in full view of those splendid mountain ranges that separate the desert from the Pacific coast.

In Pasadena, Russell and Alice Porter received and entertained us magnificently. They helped us to locate a modest bungalow where we were soon comfortably settled.

At last, we were close to the base of operations that had offered this journeyman machinist the consummate challenge of a lifetime.

Mabelle and Oscar Marshall in front of their bungalow, in Pasadena, California, 1931.

A Russell Porter drawing of Dr. George Ellery Hale in front of his 200-inch telescope.

CHAPTER XV

EXCELSIOR

Our adventure on the west coast continued over twenty years. The circle of acquaintances expanded. Natural sites — mountain, desert, ocean — offered their recurring invitations to "come and see." Family and friends came to visit us.

The compelling reason, however, for our presence in California remained the opportunity to work on the project of the 200-inch telescope. To understand the import of the challenge that carried us west, one should become acquainted with this long-term undertaking and the man who conceived and directed it, Dr. George Ellery Hale. It was his dream to bring into being that giant eye of space.

History, doubtless, will record the name of Hale in mid-twentieth century's Hall of Fame. His recognition will rest on his work as astronomer, inventor, organizer, executive, writer and editor; and not least will it be because of his sympathetic nature and his recognition of nascent good qualities in men, many of whom he assisted to attain greater heights in life's varied programs.

Dr. Hale had realized an earlier vision in the construction of the 40-inch refractor mounted at the Yerkes Observatory in Williams Bay, Wisconsin, in 1879. Succeeding dreams had become realities for him in the building of the Mount Wilson Observatory with its 60-inch and 100-inch reflectors, completed by 1919. The man's dynamism was indicated by his motto, "Make no small plans."

Richly endowed intellectually, well-trained academically, Dr. Hale exhibited unusual inventiveness in mechanical problems and in astronomical research. He enlisted the confidence and the support of resourceful men. By his magnetism he drew other intellects into cooperative effort with his great undertakings. He was

Astrophysical Building at Cal Tech, Pasadena, California. R. W. Porter in foreground. OSM photo. Portrait of R. W. Porter taken by Higgins for American Magazine, 1931.

equally successful in obtaining the best results from the mechanics upon whose skill depended the fulfillment of his objectives.

By 1930, the construction of the 200-inch telescope had been underwritten by the Rockefeller and Carnegie Foundations. The Observatory Council and the Advisory Committee were organized. A study was underway to determine the best site for the Observatory.

The Instrument Shop was the first of three earthquake-proof buildings that were erected to accommodate the colossal project on the campus of the California Institute of Technology, commonly referred to as Cal Tech. The Instrument Shop, where I was destined to work, became ready for use a month after our arrival in Pasadena.

The Astrophysical Laboratory, designed for the study of data to be collected at the Observatory, was completed in 1931. The Optical Shop, where work on the mirrors would take place, was finished the following summer. Considered one of the most dignified buildings on the campus, it adjoined the Instrument Shop by a large archway so designed to allow for the passage of the 200-inch mirror.

One of Russell Porter's first assignments, upon his arrival at Cal Tech, had been to draw plans for the Instrument Shop. Because of Dr. Hale's profound interest in mechanics, he gave particular attention to the details of this shop, its construction, equipment, and personnel. He indicated that it should be referred to as the Instrument Shop regardless of the size or the importance of work done there. As far as funds permitted he provided it with the best machinery available. This included the following: a five-ton electric traveling crane; an open-side metal planer of thirty-six-inch by forty-two-inch capacity with a twelve-foot stroke; a universal horizontal boring and milling machine; a five-foot radial drill; a heavy-duty engine lathe of a twenty-five-inch swing and bed length of fourteen feet; a twenty-inch swing lathe with a sixteen-foot bed; a sixteen-inch swing lathe with a five-foot bed; two lathes with a fourteen-inch swing and a forty-eight-inch bed; two small toolroom lathes for precision work; a small bench lathe; two small sensitive drill presses; a twenty-inch stroke shaper; a gear hobber for spur, helical and worm gear generating; and one general purpose grinder of small capacity.

The original equipment was later increased by three more lathes. The best was also provided for arc welding and oxyacetylene torch work. Every item was essential to Dr. Hale's extraordinary project.

The Instrument Shop Superintendent was Gardner W. Sherburne, known as Mickie or Mick. He had supervised the construction of the 100-inch telescope at the Mt. Wilson Observatory, but upon its completion he declined the position offered to him as night operator. Engineering problems were a kind of symphony for Mickie, but not the intricacies of astronomy.

Sherburne was well qualified to select fine tools He also recognized, among many applicants, the men competent to work with him in the Instrument Shop. Two reasons accounted for the large number of men who sought employment there. One was the growing interest in the 200-inch telescope, news of which attracted attention. The other reason derived from widespread unemployment during the depression of the early thirties.

The men hired by Sherburne came from diverse backgrounds. They brought a variety of talent to the Instrument Shop where the character of the work did not resemble that of mass production. Most of the men were journeymen machinists.

There were exceptions. For instance, Johnny Kempel of German ancestry had handled a twenty-mule team on the Mohave Desert before he worked with Sherburne at Mt. Wilson. He proved to be just the man to ride the electrically driven wheel devised by Sherburne to grind the rails for the revolving dome of the 100-inch telescope. That dome was to move with the stillness and sleekness of a cat, Sherburne warned. Later Sherburne brought Johnny to our shop where he ground the rails for the revolving dome of the 200-inch telescope.

One morning, a group of workmen stood in a huddle at the shop. As I drew near, one of them turned to me and said, "Oscar, Johnny died in his sleep last night." He had led a full life.

The following men were also among my shop associates: Ross Allen from West Virginia, an omniverous reader in his field and one of the finest mechanics I ever knew; Bill Garnett was adept in both heavy duty and fine detailed machine work; William Schultz from Indiana, a machinist of execptional patience, was precise in every detail; Art Henry, the chief pattern maker, was the best

O. S. Marshall with part of a gear-cutting attachment designed and made by him at the Instrument Shop, Cal Tech, 1931.

One of the grinding machines for 200-inch project at Cal Tech. Shop foreman Mickie Sherburne is on upper level at right.

craftsman in wood work that I have ever known, barring none; Stan Hart of Arizona, a high-grade machinist, had both natural endowment and wide experience; Don Sutton came to the Shop, having had nation-wide practice in the installation of radio equipment; John Mitchell, our welder and oxyacetylene torch expert was — unlike Longfellow's *Village Smithy* — slight of stature, but competent in his field.

Our electrician, Merritt C. Dowd, known as Jerry, was born in Connecticut. If his father had not died, Jerry might have studied medicine and become a surgeon. He became, instead, an electrical engineer during the days when the horse-drawn street-cars were converted to trolley lines. He worked on the electrical wiring for the railroad stations in both Boston and New York City.

For health reasons, Jerry had moved to California. At first, he drove trucks up the old toll road to Mt. Wilson where the 100-inch telescope was under construction. In time, he found employment as an electrical engineer and designed the electrical equipment for Mt. Wilson. When he finished that job, he planned to retire. The Astrophysical Observatory for the 200-inch telescope, however, was in need of an expert electrician. If you would like to see miles of intricate wiring with a fine cabinet for the terminals and control board, you should consider Jerry's work on that project at Palomar.

Association with these skilled individuals and the privilege of contact with men of scientific mind[1] at Cal Tech relieved me of the boredom and the frustration that had marked my machine-shop days in Springfield. Time and circumstances broke open the geode that Russell had found back in 1919. Little did either of us imagine, when he checked the base line of my Springfield Map, that both of us might one day be involved in the building of the 200-inch telescope.

By 1935, after thorough investigation of the possibilities, the authorities reached a decision on the type of equatorial mount best suited for the 200-inch telescope. Dr. Hale called it the modified yoke design. It incorporated the Porter split-ring or horseshoe design for the north-polar bearing, a device that allows, in the field of vision, for the inclusion of the north circumpolar regions. The Mt. Wilson reflector cannot reach that area because of its closed yoke design.

The 200-inch telescope comprises four outstanding features that make it unique.

First. The instrument is electrically controlled directly by the astronomer or in cooperation with the night assistant. Thus are brought into position the various auxiliary mirrors that continue the reflected image from the big mirror to focal points selected by the astronomer. These points are in addition to the first or prime focus, which is 666 inches directly over the 200-inch mirror. This is the main position where most of the photographic work will be done, although photographs may be taken at the Cassegrain focus, which is through the 40-inch hole in the 200-inch mirror.

Astronomers never seem to be in a hurry, yet time is a precious detail in their work. A maximum of 150 nights a year for effective observation is still subject to reduction by cloudy weather. A telescope is not an electronic instrument that may ignore the clouds, and the full moon limits the number of nights when the astronomer can make useful studies with his instrument.

Second. The 200-inch telescope will reach celestial objects by a dialing system for both right ascension (east-west positions) and declination (north-south positions). There will be clocks for sun time and others for star time. Due to the earth's periodic course around the sun, the stars rise about four minutes earlier each successive night. The old method of calculation required graduated circles for the right ascension and for declination.

In connection with this function of the gigantic instrument is the vast electrical equipment. A Diesel electric outfit on Palomar Mountain supplies the current for all utilities and for the astronomers' enterprises. There are 60 motors ranging from 1/12 horse power to 1/2,000 horse power. A wiring system 120 miles long includes a large network of cables, a small one for the complete functioning of the 200-inch instrument, as well as the computer system to determine star positions and to assist in the correction of errors due to refraction or to minor errors in tracking. Finally, safety provisions guard against accidents, by the prompt arresting of all movement.

Third. The 200-inch telescope has a unique lubricating system, one that reduces the usual consumption of power. Let me call the reader's attention to the mechanical features of these two great bearings: the horseshoe ring of the north polar bearing, and the south-polar bearing that is obscured from view. The latter bearing is a perfect hemisphere 84 inches in diameter. It is meticulously ground to fit its female hemisphere bearing with allowance for an oil film about three-thousandths of an inch in thickness.

The horseshoe ring is 46 feet in diameter and weighs about 155 tons. This ring rides on two sets of steel pads. They are supported in suitable pockets about sixty degrees apart on the north pier — a spacing best calculated to avoid either a wedging tendency or a disposition for the ring to escape from its track. The ring was accurately finished by grinding. The inner circle of this ring has parallel sides from the center of curvature upward to permit the required traverse of the tube in reaching the declination positions from north to south.

To avoid prohibitive sagging of this horseshoe ring when the instrument is either well over to the east or to the west for observational use, the ring was pulled together at its horns while in the grinding process — an amount calculated to maintain a true circle in the east or west position. Thus the horseshoe ring is a little elliptical in form, self-correcting in operation.

Lubrication of these polar bearings is done by a pumping system that forces a thin film of oil — about 3/1000 inch in thickness — through orifices and small channels. The system is equipped with gauges to indicate the pressure that insures proper lubrication of the bearings. This lubricating system is far superior to the best ball bearings or roller bearings known to engineers, with regard to friction. It totally eliminates vibration when the telescope is in operation. It accounts for the small horsepower electric motor that adequately drives the big telescope in its right ascension movements. This oiling system was de-

Left, Edwin P. Hubble, astronomer for Mt. Wilson. Right, Dr. John Strong, scientist at Cal Tech in the 1930's and later at Johns Hopkins.

veloped at the Westinghouse Electric and Manufacturing Company where adjustment to the weight of the 200-inch instrument presented a serious problem to the engineers.

Fourth. The mechanical provision for the support of the 200-inch mirror is of special interest. The principle, which is also applied to the auxiliary mirrors, originated with Dr. Francis Pease, Associate in Optics and Instrument Design. It doubtless resulted from his experience with the 100-inch mirror of the Mt. Wilson Observatory. That mirror suffered annoying distortion in some of its positions and Dr. Pease devised a cantilever system that he added to the existing support. His design was a platform-scale system calculated to maintain the mirror in perfect equilibrium in any of its possible positions, and free of even the slightest distortion once the supporting mechanism was adjusted.

For the 200-inch mirror there are 36 of these scales arranged concentrically from the center to the outer areas of the mirror, at its back side. Each unit carries its mathematically calculated portion of the mirror's weight. The units are secured by bolts fastened to the supporting table for the mirror. There are 36 holes in the back side of the mirror. They are 11¾ inches in diameter and of depths varying from 14½ to 16¾ inches. The deeper holes lie toward the outer edge of the mirror. Its thinnest area is the central zone, about 5½ inches thick.

Each of the 36 units has an adjustment disc head to fit the hole close to the parabolic zone of the mirror, and this head is carried by a swivel joint at the end of a long bar that extends beyond the outside or back of the mirror. At the outside end, beside other parts, are two lead circular discs for counterpoising the mirror, carefully weighed for their function, and adjustable to obtain precise equilibrium with the mirror. A member of each support unit contacts the mirror around the hole at the back side. These shoulder contacts take care of the down-thrust of the mirror, and the disc heads take care of the radial or side position. Each unit has 31 precision ball bearings, a total of 1116 sets. The unit as a whole is designed to be flexible as a rubber-jointed man, but at the same time as firm as a rock. The ideal in Dr. Pease's mind was that these units should be fully articulating mechanisms with every joint as near a knife-edge sensitivity as was possible to obtain.

In the Instrument Shop, where the precision work referred to above was achieved, a workman's talents were often challenged by the responsibility assigned to him. Two examples will illustrate.

Shop Superintendent Mickie Sherburne knew how to defeat apparent defeat. He undertook to adapt certain machines so that they would work beyond their designed capacity of performance. One day he presented just such a problem to me. He wished to avoid sending the job outside the Shop. Together we studied the possibilities. I made a suggestion that he promptly accepted. To design and make an attachment for a gear hobber, 36" in diameter that would easily handle the worm gear oversized by eight inches, was my first real job at the Shop.

In connection with Dr. Pease's original design for a mirror support, an error occurred through insufficient study of the mechanism. It was discovered by a journeyman machinist[2] who had been given an assembly-detail drawing. With it were instructions to make the whole unit and, when completed, to test it for its suitability. While his lathe was running on some of the parts, the machinist studied the workings of the mechanism in order to see how it was expected to function. In his personal search for full understanding, he considered this mechanism at several positions around the circular mirror when in a vertical plane. From his

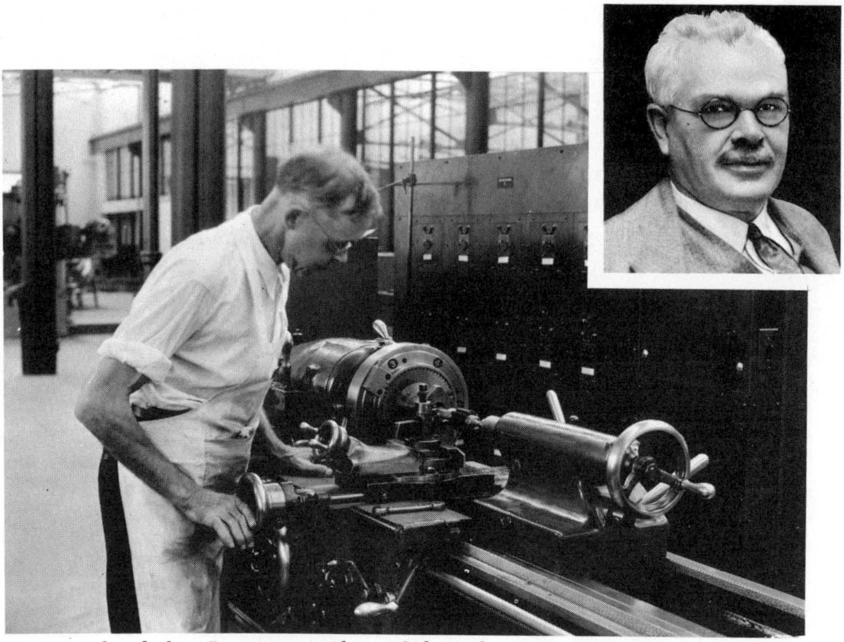

OSM at his lathe, Instrument Shop, Cal Tech. Inset, Dr. Francis G. Pease, courtesy Mt. Wilson Observatory.

analysis, he concluded that those supports above the center of the mirror would hold it back against its cell as planned, but that those below the center would push it away. Obviously this violated the purpose of the mechanism.

The machinist reported his discovery to Mickie Sherburne. He took up the problem with the engineer who had detailed Dr. Pease's design. As a result, Pease and his assistant gave the subject a thorough study, then issued a revised drawing of a mechanism that would function equally well in any position of the mirror, and the dilemma was solved.

It has been said that the heart of the 200-inch telescope is its big mirror. Marcus Brown, Chief Optician, was in charge of that mirror from the time it arrived in the rough at the Optical Shop until it was ready to leave for its installation at the Palomar Observatory. Dr. John A. Anderson, Executive Officer, gave the final word to endorse the finished job. Brown earned a place in humanity's hall of fame not to be occupied by another because of his ability to operate a fifty-ton crane that handled the mirror and its metal parts. He achieved the mirror's correct parabolic form, free of grinding or polishing evidence, in limits of one or two millionths of an inch.

Dr. John Strong[3] is another person whose name is associated with the 200-inch mirror. He conceived and directed the aluminizing methods — accomplished in a vacuum — for depositing on the mirror a film of aluminum less than a fraction of a thousandth of an inch in thickness. The mirror is removed from its vacuum chamber with a most brilliant lustre. Unlike the former silver-coating method no subsequent polishing is required, nor does it need to be renewed in less than two or three years. The former process lasted at most six months. The aluminum surface also offers sixty percent more reflecting capacity than the mirror with the silver-coating surface.

The individuals who outlined plans for the 200-inch telescope project had set a ten-year period for its completion. They aimed for the earliest possible date in view of Dr. Hale's declining health. Although he did not live to see the achievement of his dream, he knew, when he died in 1938, that it was certain of fulfillment.

In 1934, I had scrawled "1940 or bust!" across one corner of a mechanical drawing that hung above my lathe and from which I was working. World War II, however, forced suspension of all effort on the enterprise. The engineering and machine shop staff swung into action to hasten the end of hostilities by making optical devices for the armed forces. With resignation we adjusted to the setback in our time schedule for the building of the 200-inch telescope.

The Observatory Council had already selected a site for the Observatory on Mount Palomar. Russell Porter had designed the building and its dome.[4] Construction of both was in progress. In 1936, the giant pyrex disc for the 200-inch mirror, poured in Corning, New York, had traveled safely by railroad across the United States to Pasadena, California. In the Optical Shop at Cal Tech polishing of the mirror by a long grinding process was well advanced by 1941. With the outbreak of the war, the glass was covered and remained so until 1945 when the polishing process resumed.

The steel mechanism, which was designed by Cal Tech engineers to hold the 15-ton mirror, was built by the Westinghouse Electric and Manufacturing Company. In the summer of 1938, the enormous frame for the mirror and the immense horseshoe bearing with its tube were ready for shipment. They went by sea from South Philadelphia, Pennsylvania, via the Panama Canal to San Diego, California. From there they traveled by tractor trailers up the steep winding highway to the Observatory at Palomar.

As I worked on this chapter, in the fall of 1947, the mirror, patrolled with jealous care, made its final journey. The long hazardous ride by truck, from the Optical Shop at Cal Tech, ended safely at the Observatory building, an altitude of 5,500 feet. The site is located 125 miles southeast of Pasadena and fifty miles northeast of San Diego.

In December, when the installation of the mirror and the adjustment to its equatorial mount had taken place, man looked through his incredible creation at the heavens. After further corrective measures were made, the 200-inch telescope was ready for its mission in space.

June 3, 1948, marked a high point in the history of science. It was the culmination of the project with which I had been associated as a journeyman machinist for nearly twenty years. The 200-inch telescope of the Palomar Observatory was dedicated as The Hale Telescope in honor of its Founder and Director, Dr. George Ellery Hale. In the workshop of this man's dream I had found an answer to my need and to the cry of "Excelsior."

The 200-inch mirror ready to be levelled on to the grinding machine.

At left, the arrival of the 200-inch telescope disc, April, 1936, from Corning Glass Works, Corning New York. Photo by Mabelle Marshall.

Looking down upon 200-inch mirror from southwest corner of Optical Shop at Cal Tech. Photo by O. S. Marshall.

1/10-scale model of 200-inch telescope not complete. Photo by Oscar Marshall.

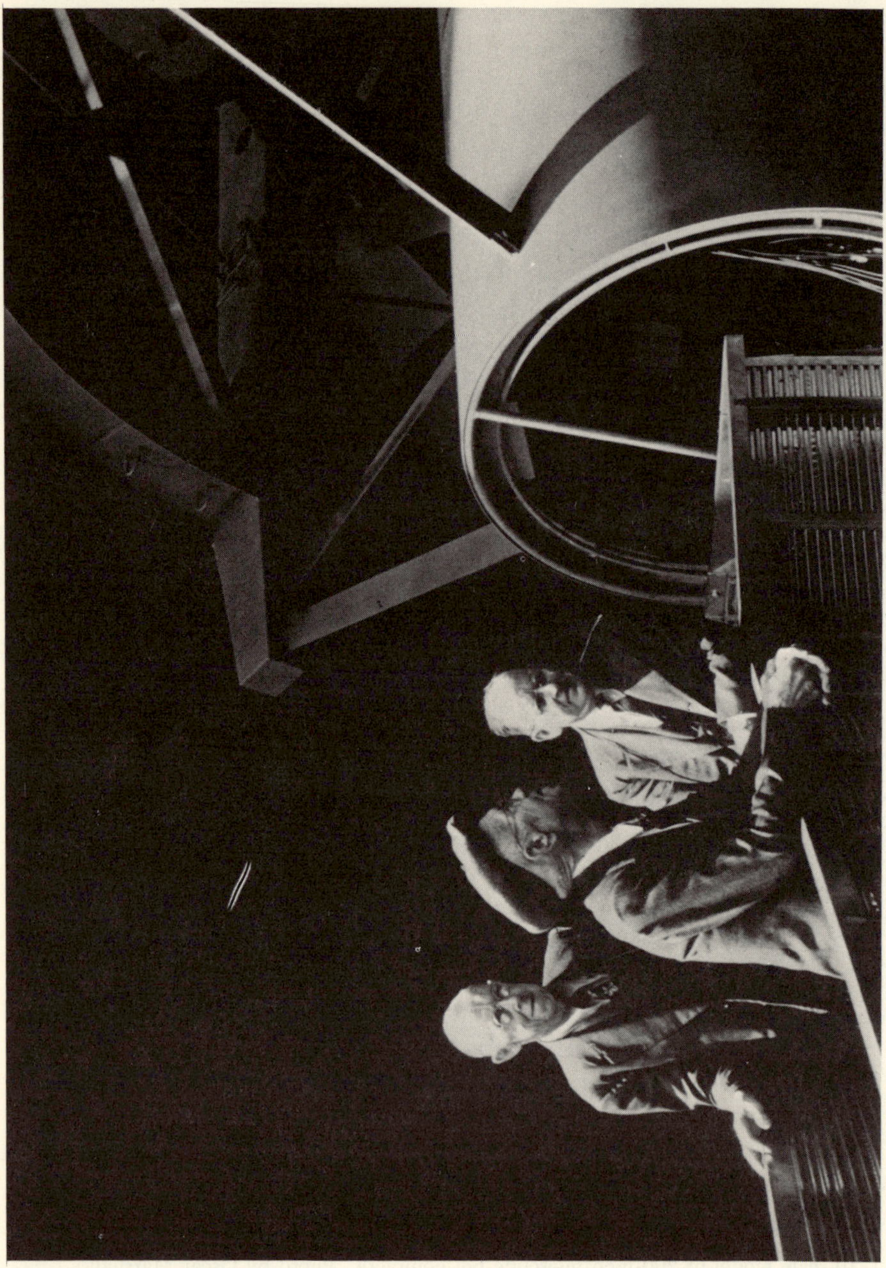

Oscar Marshall, Russell Porter, and Ernest Flanders at the prime focus of the 200-inch telescope, Palomar Observatory, March, 1947. Courtesy E. V. Flanders.

Oscar Marshall and Russell Porter at Palomar Observatory Building, 1947. Photo by E. V. Flanders.

Maxfield Parrish, the artist. Courtesy Maxfield Parrish.

Russell W. Porter at work in his office at Cal Tech.

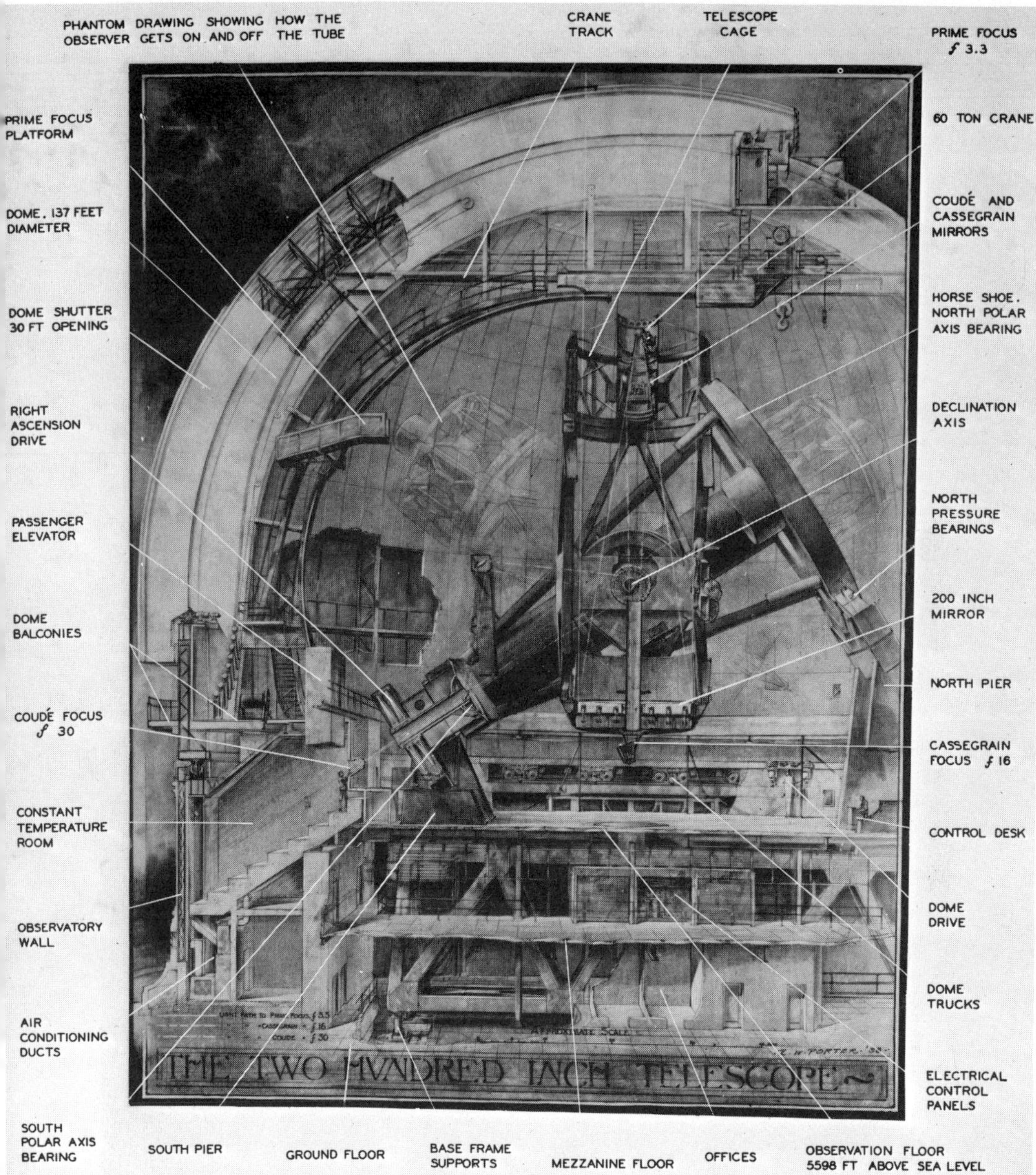

Looking west at vertical half-section of Hale Telescope.
Drawing by Russell W. Porter.

Portrait of Oscar S. Marshall in the Mojave Desert, drawn by Russell W. Porter.

Chapter XVI

AFTERGLOW FROM THE SUMMIT

Close to eighty years of climbing are behind me. From this vantage point I can survey, among other scenes, almost six decades of machine-shop life. During that time, mass production with assembly line procedures eclipsed the role of the journeyman machinist. The superhydroelectric power plant superseded the overshot water wheel. The automobile and the airplane outdistanced Orin North's horse-drawn freight wagon and the railroad. Today, television waits at everyone's door.

In 1928, the stage was set for the greatest undertaking in astrophysical science.[1] Yet, great as this achievement has been, it will not disrupt man's concept of the universe as did Galileo's conviction in the first half of the seventeenth century. To him one must ascribe the honor of first putting the telescope to astronomical use.

The 200-inch telescope is called the *giant eye of space*, but human ingenuity will devise something to outmode its particular greatness and significance. It is already prophesied. Whether or not this prophecy could be realized without the work of its forerunner, the astronomical telescope, is a question. Night has made possible, if not actual, man's knowledge of planets, stars, and galaxies. It is safe to conjecture that but for the regular hours of darkness we would still be in that more troublesome night of ignorance and superstition.

What does the future reserve for man? What will the giant eye of his big telescope reveal to him? Is man about to unlock mysteries that the planet Mars has thus far withheld from him? Who will discover the answers?

From this summit in time on planet Earth, I view the years ahead with deep appreciation for the gifts which they offer: (1) the joy of finding life continually meaningful, (2) the joy of maintaining close relationships with family and with friends, and (3) the joy of exploring a number of California's superlative features.

In addition to the satisfaction found in my particular work on the 200-inch telescope at the Instrument Shop in Pasadena, I have had time and occasion to pursue interests in photography, in pastels, in oil painting, and in reading. An insatiable desire to write has found expression, not only in an extensive correspondence with widely scattered friends, relatives and immediate family, but also in the writing of various articles for publication.[2]

In spite of time and distance that separate us from friends and loved ones, we maintain close ties with them. We have done this by a two-fold process: a lively correspondence, which both Mabelle and I enjoy, and personal visits. Not only have we made return trips to the eastern States, but we have received scores of loved ones in our Pasadena home.

Mabelle and Oscar Marshall in desert, bidding farewell to daughter Eva and family as they return East, January, 1941. OSM.

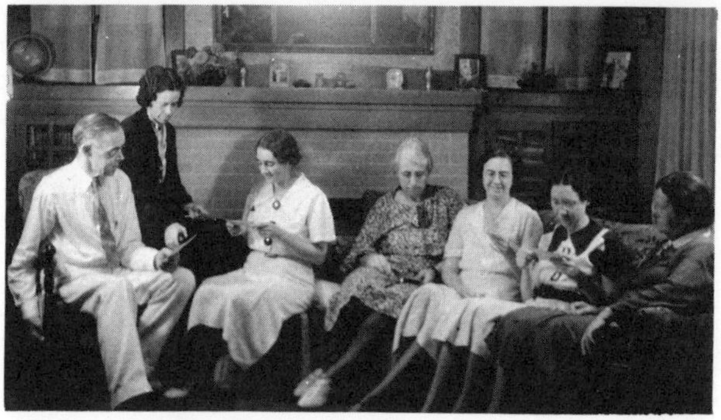

Top left: Kathryn Dougles, Loyal Wright, Jr., Paul Douglas, Oscar Marshall, Eva Douglas, John Douglas. Top right: Elmer Douglas, Mabelle Marshall, Eva Douglas, Paul and Kathryn, Christmas, 1940. Center: Oscar M., Candace Marshall, Helen Hageman, Adaline Hagaman, Elizabeth Hageman and friend, Mabelle Marshall, 1930's. Lower left: George Marshall with grandson George Howard and Oscar Marshall, August, 1946, at 1114 E. Orange Grove Ave. Lower right: Oscar Marshall, Virginia Dugan, daughter of George M., Mabelle and Candace M., George Marshall, 1935, Pasadena California.

I find it impossible to dissociate these joys from that of exploring California's superlatives, as I like to call them. Among the exciting discoveries have been her deserts and Pacific shoreline, her national parks and mountains, as well as her Franciscan Missions. We have explored many of these with our visitors. We have written about these occasions and we have made them the subject of photographs and pictures on canvas.

As I linger here on the summit in the glow of the recent years, let me sketch a few of these scenes as I recall them.

First, the desert. We have set up camp there both when the heat was good for arthritis and when the ground was so frozen that we had to chip out our fuel for the campfire. Snug in our sleeping bags, we lay under the stars — stars like diamonds almost within finger reach. Occasionally, a star or two would skim across the night-blue sky, leave an angel trail fanning out in space and then disappear. Night on the desert was never too long. We make an exception, however, to one disturbance: the cloudburst, a terrestrial beast to be avoided whenever possible.

The Mohave Desert was a favorite retreat for us and our visitors. One evening, several of us sat around the warmth of the campfire after the desert sun had slipped behind the range of austere mountains. The Joshua tree logs, because of their punk-like quality, burned slowly. An enormous bird, the western horned owl, swooped above us, his wide-spread wings outdoing the night breezes. He was pursuing a rabbit for his evening meal.

Someone in the group inquired about the Joshua trees with their bayonet-like leaves. We explained that once in four or five years they put forth luxuriant blossoms in bunches the size of a large white goose. One tree may display twenty clusters of creamy-white flowers each blossom of which matures into a pod the size of a goose egg. According to botanists, the Joshua tree is related to the lily family.

When the Mormons trekked through the desert in the late 1840's, the silhouette of these tree limbs reminded them of the uplifted arms of the Biblical Joshua. Hence, these Latter Day Saints called them Joshua trees.

The Mohave Desert offered also to Russell Porter and to me a special haven to which we escaped at various intervals to indulge our hunger for solitude. We sketched and painted. I photographed a gem in nature or a fine moment in time. We roughed it around the campfire as we had done back in Vermont.

Mabelle and I remember those years when winter rains drenched the desert. Spring flowers then quietly spread their royal colors, like a Persian carpet, over the apparently barren expanse of desert. City dwellers, like ourselves, drove many miles to see the miracle: California poppies of velvet gold, steeples of blue lupine, verbena's variegated shades of pink and red. Such were the contrasts in the desert, studed elsewhere with cactus families and Joshua trees.

Through the years I have heard the Desert's plea:

> I call you Friend, now. I come to you from out the ages. I tremble lest you will not give me a place by your hearthstone.
>
> My record is bad, all the crimes of Nature are justly charged against me. My accusers are legion, the evidence overwhelms me. I am called treacherous, deceptive, and unpitying. Masked by gorgeous colors and intriguing forms and concealing fabulous wealth, my chief delight is death, death by slow torture, attrition, while I gloat over my victims. Men, with their wives and children, and with their animals, have arrived at my borders, wonderingly, hesitatingly, lured by the rainbow of colors, by the minarets rising from out my mysterious seas, my illusory mirages.

Photograph of OSM's painting of a Joshua tree.

These men, buckling on their assurance, have then set forth to explore the unknown only to go down to a miserable end, death by piece-meal, by slow torture of heat, hunger, thirst. Heat they may endure, and hunger for a time, but thirst gets them eventually. Food of a sort, and water too, are available if only man's wits would come to his aid.

Throughout untold ages I have lived riotously with myself. My drifting sands, my dust devils, my heroic sand storms, which smother every living thing, these have I adored. My boulder-strewn watercourses I have loved, where cloudbursts have joined me in wreaking desolution such as no human army could accomplish with its toy-engineering devices.

My animal life is unique, and rare, venomous to a degree, though ages of selection, adaptation, and specialization. Diminutive in size, artfully deceptive in form and color, with lightning unexpectancy, each one executes its deadly work.

My flora, admirably suited to my purposes, is equipped with spines, barbs, and bayonets for every occasion, ready to irritate, entangle, and impale its victims. Apparently it is the most useless vegetation for which none but a genius may discover a scintilla of good. The color and variety of my flower gardens beggar description and defy man's childish efforts to imitate. Lonely canyons and desert floor, jagged cliffs and storm-strewn water courses, these constitute my rock and flower gardens. Uncreated are the pastel colors of human artists that can approach my offhand works.

My mineral and chemical possessions have proved to be the real source of temptation that has lured man upon the road of exhaustion and death. Gold, silver and precious gems, these man is determined to have. I have distributed them, broadcast, in rocky defiles and in vast lonely stretches. There either cloud-bursts or storms of whirling sands were sure to leave, sooner or later, nothing of man or his animals but bleaching skeletons.

Both the earth and the sky belong to my realm. The earth is my floor; the sky is my canopy. Ah, the Night! Then the stars come near me, so near that they seem to

Left, yucca in blossom, Aroyo Seco Canyon. Picture taken by Elizabeth Hageman. OSM inspects the blossom. Right, "Pals," (The author makes friends with a burro), "Early California Days." Photo by OSM.

Russell Porter and OSM by their campfire in the Mojave Desert, February 28, 1931.

hush the whispers of earth. The fierce blaze of day is turned to balmy coolness. Despite my native tastes, I succumb for the moment to that effeminate spell wafted earthward by Pleiades . . . then catch myself in time to continue my age-old assaults upon intruders of my realm.

Thus for ages was I an effectual barrier to man, a deadly passage for all who essayed to reach that Land of Gold on the shores of the Pacific. I conceived my task as simple, certain, permanent. But, alas, I am undone, outwitted, conquered, humiliated. And the victor? None other than my age-old victim, he with whom I thought it so simple to toy.

Man has become my conqueror, and to Man I submit. An ignominious future awaits me . . . unless I can make my peace with him. It is fortunate for me that Man is naturally a forgiving and a forgetting creature. I shall seek his pardon, offer him my best services in cooperation with his aims and his efforts to subdue Nature.

From the captive spell of the desert, I now turn my gaze to California's shoreline that also captivated Mabelle and me on many occasions. At La Jolla and at Laguna, nature's spectacular scenes tempted the would-be artist within me, painter and photographer alike. The violent breakers and the calm of low tide held us spellbound. They quenched our thirst for refreshment not found in the confines of the city.

Above both the desert and the shoreline of the ocean, I lift my eyes, remembering California's lofty grandeur in Sequoia and Yosemite National Parks. At the former, Nature gave us relaxation through her immense areas of solitude and her vistas of scenic beauty. We camped, we read, we wrote. We took pictures, and I sketched the outlines of scenes to finish later with oil or pastel.

In 1932 we stood at the base of the giant General Sherman Tree, one of Nature's oldest living organisms. I felt that some ritual might have prepared us to meet this monarch, this living embodiment of an ancient past. How can we face the uncertain significance of our short span of life on the same planet with him?

In May of 1934 we explored parts of Yosemite National Park, that glacier-created wonderland[3] of California, another of her superlatives. The weather gave us such a fine reception that our

Oscar S. Marshall painting in the desert.

Son-in-Law Elmer H. Douglas in Panamint Valley, California.
Photo by OSM, 1940.

5 x 7 black and white photographs bear witness to that happy excursion.

One of my interests in Yosemite was the massive formation sometimes known as the Three Brothers, giant buttresses of granite. Their facades bear scars of the ancient conflict in which they withstood the movement of glaciers twenty thousand years ago. At the time of our visit, the Brothers greeted the morning sun. Their rugged forms stood mirrored in the placid surface of the Merced River four thousand feet below the highest Brother.

What particular interest did I have in these three peaks? As a youth, I had found a small halftone reproduction of these so-called Three Brothers, but no indication of their location. I used the print as a copy to make an oil painting, which hung for several years in our Springfield home. I never suspected that I would one day stand in Yosemite and gaze at the faces I had painted so long ago.

From these eternal wonders of Nature in California, I turn to one of her man-made creations that was started in 1769. It is El Camino Real, the Royal Road along which Franciscan Missions were established from San Diego to San Francisco. Through these missions, Fra Junipero Serra, a Franciscan Monk from the Isle of Majorca, hoped to civilize the Indians and to convert them to Christianity.

We have thought it worthwhile to become acquainted with a number of the Missions. Those at San Juan Capistrano, San Gabriel and San Fernando we visited a number of times.

Oscar and Mabelle with daughter Candace at General Sherman, Sequoia National Park, California.

OSM and Mabelle at Three Brothers, in Yosemite Valley.

A view of waterfalls in Yosemite.

Left, Oscar Marshall in San Juan Capistrano, 1932. Photo taken by Candace Marshall. Right, Stone stairway to choir loft in Old San Gabriel Mission, built in 1771, at San Gabriel, California, by early missionary padres to Christionize the Indians. Entrance to Auditorium is at left. Photograph by OSM.

San Juan Capistrano Mission. Photograph by Oscar Marshall, 1933.

From this summit of the years, I see the faces of many persons who have meant much to me. I would include their names if space permitted.

It is time to turn the eyepiece closer to home for a look at a more recent scene.

During my vacation in August, 1946, we traveled by Pullman train from Pasadena to see our families and friends who still lived in the East. We also planned to attend an important event.

The first post-World War II Convention of the Amateur Telescope Makers was scheduled to take place that month at Stellafane, near Springfield, Vermont. The chief attraction for the three hundred amateur telescope makers was the presence of Russell Porter. He had come with his wife Alice for the occasion. Mabelle and I appeared unannounced. (Our friends seemed to think that this was a bit like rising from the dead.)

Brother George and his grandson George Howard were also there. Our hands were still warm from the farewells exchanged but a short time previously, after visits at Yosemite and Mount Palomar. Yes, from Mount Palomar back to Stellafane and return, this journeyman machinist would soon travel full circle to keep faith for two more years with his peers at that site where perfection was paramount.

On that particular week-end, before leaving Stellafane, "*Temple to the Stars*," I gazed at length across the familiar valley to the Green Mountains. Against the distant horizon, Mount Ascutney lay bathed in the lavender afterglow of an August sunset and I thought,

> "There's a divinity that shapes our ends,
> Rough-hew them how we will."[4]

At Laguna Beach, California, July, 1938. Photo by OSM.

Left to right, Candace, Beulah, and Eva Marshall, circa 1918.

EPILOGUE

Oscar Marshall, father of three daughters, Eva, Candace, and Beulah, gave us, among other gifts, those of love, encouragement, understanding, and time. Reliving our childhood years, as I worked over this manuscript, has brought into relief the stature of the sensitive person with whom we lived and grew into adulthood almost unaware of his humble worth.

Sunday afternoon, while Mother enjoyed a period of uninterrupted rest, the one whom we fondly called Daddy-man took us for a walk in Hartness Park. There he taught us to recognize the checkerberry, tiny red fruit of the wintergreen plant. We learned not to say peroxide when we meant the witch hazel bush. Seated by the edge of a tumbling stream, our nature-loving father challenged us to find just the right adjective that would describe the water's changing moods.

It took us quite some time one Sunday afternoon to discover that our escort was playing the role of the hoot owl whose call he imitated with cupped hands. We anticipated those outings when he taught us to do the same. He also taught us to identify a variety of birch trees in Hartness Park, the songs and markings of its many birds, and the names of its modest wild flowers that we looked for every spring.

Wintertime found us with homemade skis and scooters. Snowshoeing was great fun when the outdoor-hungry parent took time to go along with us over the fields of fresh snow dotted with rabbit tracks. Sometimes, when the weather was fair, he would leave the house early Sunday morning for a long hike or a bicycle ride by himself, but he returned in time for Mother's delicious pancake-and-sausage breakfast with Vermont maple syrup. If he found us

Oscar Marshall with his three daughters. At rear is Eva, with Candace and Beulah in foreground and right, respectively. Hartness Park, Springfield, Vermont.

The back of Hartness Park, Springfield, Vermont.

girls in bed, he would play a favorite phonograph record, Harry Lauder's Breakfast in Bed on Sunday Morning. It never failed to get a rise out of us.

A deeply religious person, this man of varied interests was an active member of the Methodist Church. I remember, with a bit of nostalgia, hearing him sing with the congregation, his deep bass voice mellow and true. Forward-looking and tolerant, he encouraged us to cultivate similar attitudes. He was a fun-loving, jovial individual who savored a good story or a healthy joke. Several years slipped by before we discovered that Santa Claus, who came regularly on Christmas Eve to the church program, was none other than our own father having a hearty time with everyone.

My sisters and I cherish the memory of family picnics. Our friends were often included on these outings to hilltops and mountain peaks.

Prominent among our memories stands *Antares*, the reflecting telescope that we watched emerge in our own backyard. Its enthusiastic master-builder, our Daddy OSM, invited us in due time to "come and see." Sometimes it would be in broad daylight to have a peek at Venus. On winter evenings we bundled up at his invitation to look at the craters of the moon or to observe the changing positions of Jupiter's satellites. Little did any of us dream of man's exploits on the surface of the moon or of his Viking ventures on Mars, which would occur fifty years later.

In our teens, my sisters and I found it a bit difficult to choose a vocation. Continuing our education beyond high school seemed very uncertain. Our parents, however, stood by us with patience and wisdom. Thanks to their guidance and that of our friends and relatives, we unravelled the threads that led to our careers.

During this period of struggling to find our way, we became aware that our father's own career bore the scars of shattered dreams, but the enduring impression was one of resilience without self-pity or complaint. Having sensed a measure of his life-long battle with fractured hopes, we joined Mother in gratitude when he received his reward as an expert journeyman machinist, namely, the invitation to become part of the 200-inch telescope project in California.

While our parents broadened their horizons with new opportunities on the west coast, my sisters and I did the same in other settings. In 1931, I married Elmer H. Douglas of Kingston, New York. We have a daughter and two sons. Beulah married Loyal S. Wright, Jr., of Chazy, New York, in 1934, and they have two daughters. In 1940, Candace married Joseph B. Hartshorn of Somerville, New Jersey. They also have two daughters.

In July, 1952, just before returning to our missionary assignment in Tunisia and Algeria, I acknowledged with misgivings the depth of my father's prolonged suffering, both physical and emotional. I knew that our farewells would be final.

I asked him, "Do you still believe in God and in His goodness for man and the universe?"

I coveted his affirmative answer. It came with resounding conviction, "I do, never fear, my dear girl, and I believe that He has the universe in hand."

Then we said goodbye.

Recently, I have had occasion to reread many of the letters received from my father, who wrote regularly between the years 1930 and 1952. Without intent, they reveal the rhythm of his inquisitive mind, the limitless reach of his spirit, the deep concern of his loving heart. In 1936, he wrote the following lines, at Christmas, to Ernest Flanders, his friend of almost seventy years:

"In our younger days, while we knew that eventually we should be old like our parents, still the flush of youth would not permit the real sensing of the inevitable fact. To grow old charmingly is an art, not always, however, an attainment. The best aid in accomplishing this desirable goal is confidence in the future, a trust in an Eternal Goodness that will repair all our faults, all our shortcomings. I want to live always — forever — somewhere — somehow. Let us believe that we shall, let us hope that we shall, let us live as though we were sure that we shall — this is the wish and the aim for you and for us. Oscar and Mabelle."

With this expression of faith that remained undaunted to the end of his life, we bring to a close the account of our journeyman machinist en route to the stars.[1]

Photograph of a painting of Stellafane by OSM.

Jan Lippershey, inventor of the principle of the telescope in 1609 in Middleburg, Holland.

NOTES

Chapter I

1. Alden Velona Marshall was born September 30, 1838, into a large family of farmers, in Lempster, New Hampshire.
2. Eaton B. Marshall married Edith Humes. They had two daughters: Ruth M. Bjornsen (Mrs. Henry) and Marian M. Hall (Mrs. Horton).
3. George G. Marshall married Mary Batchelder. They had two daughters: Isabell M. Howard (Mrs. Phillip J.) and Virginia M. Dugan (Mrs. William E.)
4. Velona A. Marshall married Kathrine Hulbird.
5. Alden V. Marshall's preaching appointments included Mendon Vermont; Moriah, Redford, Schroon Lake, Knox, New York.

(Unpublished MS)

Chapter III

1. For the story of Jones and Lamson Machine Company, see the following: *Precision Valley*, The Machine Tool Companies of Springfield, Vermont, by Wayne G. Broehl, Jr., Prentice Hall Inc. 1959, Englewood Cliffs, New Jersey. (Ed.)
2. Gilman did buy the following modern machines: two Reed Engine Lathes, Putnam Lathes and a Jones and Lamson lathe later discontinued. For his Gilman-built gear-cutter machines, he bought a 30" index plate. Before I completed my apprenticeship, Gilman obtained a Mushet tool and one or two semi-high-speed steel tools. The elder Gilman created certain taps, special reamers, a set of screw-cutting dies and a machine for bolt threading. (Unpublished MS)
3. The Gilmans' son, Wilbert, graduated from Boston University and took a two-year course at Worcester Polytechnic Institute before he became a junior member of the new firm, Gilman and Son.

   The Gilmans' daughter, Gertrude also graduated from Boston University. She was a missionary in China, and retired in Pasadena. Ibid.

Chapter IV

1. During my apprenticeship and for some time afterward, Mother and I lived across the street from the Lockwoods, Eugene and Edna, and their mother. They were among my first and most intimate friends. In this community I was making a new start in an effort to blot out my disheveled background. It was the reaction of a sensitive person, drifting in the shadow of three successful brothers. As a result of my new associations I became absorbed in reading historical, sociological and philosophical material. My restlessness with shop life increased until I decided to do something about it. (Unpublished MS)

Chapter VI

1. "Springfield industries for the most part were among the first in the country to adopt shorter working hours." *Springfield Reporter,* July 26, 1923, from an article entitled "Springfield Then and Now," by O. S. Marshall. (Ed.)
2. Mother Marshall died at sixty years of age, at the home of her son George where her four sons were present. (Unpub. MS)
3. In June, 1974, Bernard Lashua was the photographer for the fiftieth reunion of the Springfield High School Class of 1924 (that of the Editor).
4. Gilman's business was bought by W. W. Slack and H. K. Parkman in 1907. The shop building was bought by the Fellows Gear Shaper Company in 1939.
5. Experience teaches the stupid. (Latin)

Chapter VII

1. Frank Hill was Shop Superintendent at Jones and Lamson's 1904-1909. (Ed.)
2. Ben became a long-time friend of our family. Years later, in the absence of my parents, Ben and his wife were with me when commissioned for overseas missionary service. Their daughter-in-law, Mrs. Gladys Billings, had played an effective role in the planning of my career. (Ed.)
3. The Hartness Cross-Sliding Geared Head Turret Lathe appeared in 1903. With this machine, the Jones and Lamson Machine Company became a major participant in the field of mass production. Hartness, a man of high ethical standards, was recognized for his inventive skill, his executive ability and his gift for discerning the potential merit of a mechanical device. (Unpublished MS)
4. Charlie Johnson served his journeyman machinist apprenticeship at Parks and Woolson Machine Company, in Springfield, Vermont. In October, 1888, with W. D. Woolson and Myron C. Whitcomb, Johnson supervised the actual transfer of the old Jones and Lamson property from Windsor, Vermont, to Springfield. (Unpublished MS)
5. Fred P. Lovejoy, born in Andover, Vermont, May 9, 1874, died an accidental death in 1939. A machinist and draftsman with marked ability for designing tools, he served Jones and Lamson's from 1895 until 1916. Hartness then encouraged him, as he had done others, to establish his own plant, which became the Lovejoy Tool Company. Ibid.

6. In 1910, Jones and Lamson's hired young Ralph Flanders. Hartness, who recognized Flanders' genius for handling mechanical problems, put him in charge of manufacturing the Fay Automatic Lathe acquired from a Philadelphia company in 1912. Events moved swiftly for Ralph Flanders, whose talents carried him to the top of the Jones and Lamson Company. He became its President in 1933 when Mr. Hartness retired. Ibid.

Chapter VIII

1. The author contributed articles of technical nature to the following magazines: *Factory and Mill, Machinery, American Machinist.* (Ed.)
2. The author is indebted to Mr. Alfred R. Stock, Librarian of the British Institution of Mechanical Engineers, Storey's Gate, St. James Park, London, S. W. Facts about Bessemer, Koransson and Mushet were obtained from *The Memoirs and Proceedings of the Manchester Literary and Philosophical Society,* England, Vol. 57, (1912-1913), article therein by Ernest F. Lange. (Unpublished MS)
3. Charles P. Steinmetz (1865-1923) was German-born. He came to the United States in 1889 when man's understanding of electricity was in its infancy. He was a research engineer for the General Electric Company, Schenectady, New York. (Unpublished MS)
4. Early in World War I, Mr. Charles Norton took me through the Norton Works in Worcester, Massachusetts. Well acquainted with the Norton Grinder, two of which I had in my department, I was grateful to meet this pioneer in the field of grinders. Ibid.
5. William Gleason's name is well known in the industry that grew out of his thrift and genius, The Gleason Works, which are outstanding in bevel-gear machine design. They celebrated their eightieth anniversary in 1945. Ibid.
6. William E. Sykes trained at Leeds University in England. In 1923, he joined the Farrel-Birmingham Company, Buffalo, New York. In 1934, he was awarded the Edward Longstreth Medal by the Franklin Institute of the State of Pennsylvania, in recognition of the Sykes gear generator, continuous tooth herringbone gear and other work of scientific value in the field of gear design. (T. V. Busk, Farrel-Birmingham Company, Ansonia, Connecticut. Obituary notice, June 26, 1943.) Ibid.

Chapter IX

1. Tragedy entered our family when Mabelle's father Edward Maranville was stricken with pneumonia and died in June, 1914. (Unpublished MS)
2. Among the articles written by the author for *The Springfield Reporter* were the following: (1) February 6, 1914, "Highest Peak in Town." Oscar Marshall settles an old Controversy: Camp Hill 1538 feet, Mt. Ephraim 1427 feet. (2) January 11, 1915, "Our Maps and Atlases." A Review of the New Face for Springfield. (Ed.)
3. If by some cosmic trick I found myself in Middleburgh, ancient capital of Zeeland, Holland, I would visit its museum to see Hans Lippershey's spectacle telescope. From the invention of that obscure spectacle maker, Galileo

found his inspiration for making a real telescope that revealed to him the moons of Jupiter. (Unpublished MS)

4. Mr. Bernard, by some called a hermit, lived a solitary life remote from town. My sisters and I considered it quite a lark to ride with our parents to his rustic place and watch him boil down sap into golden maple syrup. One Saturday afternoon, Louise Prevost, later Mrs. C. A. Seale, and I hired a pony and two-wheel cart. We drove several miles over country roads to deliver one of Mother's apple pies to the appreciative gentleman. (Ed.)
5. Following are titles of two of the author's humorous articles that appeared in the Jones and Lamson Column of the *Springfield Reporter*:
   (1) *Great Convention of Jones and Lamson Spirits* at which Mr Balaam Sack Jas of Thistledown spoke on Social Hydrophobia. March 3, 1921
   (2) *A Rope Story,* an account of Hartness and Porter with W. D Woolson on a strange automobile trip. (October 23, 1919) (Ed.)
6. Engineers whose reputation figured prominently in this period at Jones and Lamson's: James Hartness, Ralph E. Flanders, George Perry, John E. Lovely (Unpublished MS)
7. "The Telescope as a Measuring Instrument," by Oscar S. Marshall, article published in *Popular Astronomy,* March 1922, Goodsell Observatory, Carleton College, Northfield, Minnesota. (Ibid)

Chapter X

1. *Russell W. Porter: Arctic Explorer, Artist, Telescope Maker* by Berton Willard, 1976 The Bond Wheelwright Company, Freeport, Maine 04032 (Ed.)
2. *The Arctic Diary of Russell William Porter*, edited by Herman Friis, 1976 University Press of Virginia, Charlottesville, Virginia 22903 (Ed.)

Chapter XI

1. . . . "The evening came when the eager group started grinding. Barrels, filled with stone, were all ready in place for each member. Each had two plate glass discs of the same size, from four to nine inches in diameter. One glass, to be the tool, was stuck to the barrel-head with pitch, and to the other — the speculum — a wood handle was also stuck with pitch. Several grades of carborundum and plenty of water were used while the operator rubbed the upper disc back and forth over the under disc, and slowly walked around the barrel, keeping the strokes of equal length.

   Mr. Porter said that that particular corner of the basement had been chosen because it was absolutely necessary to work in a place where the 24-hour temperature was constant, so that the expansion and contraction of the glass wouldn't be too great. There was also pure running spring water. This was important because any grit left on the glass would cause pits when the grinding resumed. Mr. Porter warned against letting abrasives fly around. Care to be clean, lightness of touch, and patience were in order . . ." From the *Springfield Reporter,* August 15, 1956, by Gladys Piper Skinner of Windsor, Vermont, member of the original group of the Springfield Telescope Makers. (Ed.)

2. The Springfield group of amateur telescope makers adopted the newly coined word *Stellarfane* as the name for its home on Breezy Hill. The suffix of the name *Newfane* had stimulated my curiosity. I found that *fane* of Anglo-Saxon origin meant sanctuary, shrine or temple. It occurred to me that here was a suitable word by which to 'portize' our amateur headquarters. I proposed my coined name Stellarfane to Porter and Fullam who seized upon it cordially. The subsequent misspelling wherein the *r* was dropped seems less appropriate because it emasculates the form . . . In generous spirit, however, I accept the existing form. (Unpublished MS)

   The words "The Heavens Declare the Glory of God," (Psalm 19:1 King James Version of the Bible), frame the front gable of Stellafane and mark the presence of this *Temple-to-the-Stars.* (Ed.)
3. I continued to write articles for the *Springfield Reporter,* for *Machinery* and for *American Machinist.* (Unpublished MS)
4. *Antares*, named for the red star in the constellation Scorpio, was the first celestial object the author found by calculation on his eight-inch reflector. (Ed.)
5. *Antares,* my father's eight-inch reflector, had not been in use for several years. At the Jones and Lamson Machine Shop, in 1976, it was restored under the careful supervision of Mr. Fay Kingsbury who remembered its builder. The instrument stands in the Hartness-Porter Museum operated by the Springfield Telescope Makers, which is located on the grounds of the Hartness House in Springfield, Vermont. A smaller instrument, built for his brother George Marshall, was restored by Berton Willard, and is also there. (Ed.)
6. In answer to my father's query about his first meeting with Russell Porter, Mr. Albert Ingalls replied as follows in a letter dated November 21, 1947: "Dear Oscar: . . . I was in New York Public Library, waiting for delivery of a book. While waiting, I picked up a book that lay on a desk . . . a bound volume of *Popular Astronomy,* Volume 31. I opened it idly and it opened to page 153, March 1923. Article on that page "The Telescope Makers of Springfield, Vermont," by Russell W. Porter. Before my requested book arrived, I had read that article and taken the author's address . . . Later he was in New York and we spent an evening at Brown's Chop House, chewing the rag about telescopes. That was the famous occasion when, in answer to my question about how you ground a mirror — the stroke — what it was like, he inverted one saucer over another, stood up, and went deliberately through the motions. He stopped now and then to explain. After a minute or two, the whole place had quit eating and was watching the funny man doing what must have looked like a sort of washboard act at the table, in a Broadway restaurant at or just before theater-going time . . ."

   In 1925, not long after that encounter, Ingalls went to Stellafane. In the spring of 1926, Porter wrote two articles for *The Scientific American.* Ingalls continues in his letter to the author, "These articles brought 368 requests for more, so we published *Amateur Telescope Makers* in 1926 and the fat was in the fire . . ." (Ed.)
7. *The Scientific American,* May 1943. "A Hobby Goes to War," by Albert G. Ingalls. November 1944, "Teletopics," page 239, by Albert G. Ingalls (Ed.)

Chapter XII

1. There is a slight variation in the detail of this story that my father recounted elsewhere, but the outcome was the same. (Ed.)
2. "Mount Ascutney as Seen by a Scientist," by Oscar S. Marshall, in the *Springfield Reporter* August 1, 1929. (Ed.)
3. Mr. James Hartness was Governor of Vermont for one two-year term, 1920-1922. (Ed.)
4. Russell told me years afterward that as they all looked over the large-scale design proposed for the giant reflector telescope, they noticed a singular similarity in the assembly drawing to that of the Porter Garden Telescope. Russell had made a set of drawings for a mammoth reflecting telescope remarkably like the set that Dr. Pease showed at the conference that day in the Governor's den. Russell had sensed, however, that this instance in his personal experience was not a solitary happening in mental processes that related to inventions. (Unpublished MS)
5. "A New Form of Mounting for Large Reflectors," by Russell W. Porter *Popular Astronomy* March 1918, Volume XXVI No. 3 Page 147 Published by Goodsell Observatory of Carleton College, Northfield, Minnesota. (Unpublished MS)
6. John M. Pierce had contributed immeasurably to the growth of STM by establishing a clearing house for materials and practical information to assist amateur astronomers in their hobby. Ibid

Chapter XIII

1. "Secrets of the Universe in Which We Live," *Springfield Reporter* December 13,1928 (Ed.)
   "Proportional Form-Turning Machine," (Thomas Blanchard's machine used in 1830.) *Machinery* December 1928 (Ed.)
   The Vermonter Vol. 33, 1928, pp. 118-122 "Russell W. Porter." (Ed.)
2. Ernest Flanders was the brother of Ralph Flanders, son-in-law of Mrs. Hartness. (Ed.)
3. As so often happens in groups such as ours had been, there were those who wanted further discussion of related subjects. We held some extra sessions in an effort to give at least a partial understanding of questions raised. (Unpublished MS)
4. See *Sky and Telescope* November, 1975, Vol. 50, No. 5, "Gleanings for ATM's," by R. E. Cox and R. W. Sinnot (Old Telescopes at Hartness House) (Ed.)

Chapter XIV

1. With chin up and a smile, our parents faced a difficult part of the challenge in their new venture. They left knowing that two of us remained with unfinished preparations and scheduled events that they would have to forego. We all shared the strain of this new beginning that to a certain degree pulled us apart. (Ed.)

2. We had hoped to meet Clyde W. Tombaugh, an Amatuer Telescope Maker, who had, in January, 1930, spotted Pluto, ninth planet of our system. He had done so on a photographic film of the Lowell Observatory. Tombaugh, however, was at a football game that afternoon. (Unpublished MS)

Chapter XV

1. Following are some of the men of science whose names I remember hearing in conversation and recall reading about in my father's letters: George E. Hale, Robert A. Millikan, Arthur A. Noyes, Albert Einstein, Ira S. Bowen, Edwin P. Hubble, Fritz Zwicky, John Strong, Francis G. Pease, John A. Anderson, Bruce H. Rule, and Russell W. Porter, a personal friend of the family from the early 1920's in Springfield, Vermont. (Ed.)
2. Presumed by reliable sources to be the author himself. (Ed.)
3. I took a photograph of John with his first bell jar or vacuum tank for aluminizing small pieces of glass as an experiment preliminary to what he hoped to accomplish when the 200-inch mirror would be ready for its final surface treatment. His tests paved the way for his successful achievement of the aluminizing process. (Unpublished MS)
4. Through the dedicated concern of Horace Brown, artist of Springfield, Vermont, some of Russell W. Porter's drawings of the 200-inch telescope have been preserved in Springfield, Vermont. Following is a copy of Mr. Brown's letter to the author, together with a copy of the appreciation written by Maxfield Parrish. (Unpublished MS)

December 19, 1938
Springfield, Vt.

Dear Oscar:

I am enclosing herewith a copy of Maxfield Parrish's appreciation of Russell's drawings of the 200-inch telescope.

As you probably know, I got a set of these for the local Springfield Library, have had them framed, have obtained from Russell an autobiography to go with them, which shows his association with Springfield and his accomplishment in later life. This appreciation by Parrish, also to be framed, will indicate to the people of Springfield how wonderful these drawings are.

I have an extra copy of this appreciation for Stellafane and one for *The Scientific American.*

Sincerely yours,
(signed) Horace Brown

The appreciation by Maxfield Parrish:

Every once in a while a body would like to go out and tell the world something. I had this feeling when looking at the remarkable drawings by Russell Porter for the great 200-inch telescope now in course of construction on top of Palomar Mountain in California.

If these drawings had been made from the telescope and its machinery after it had been erected they would have been of exceptional excellence, giving an uncanny sense of reality, with shadows accurately cast, well nigh perfect perspective: but to think that any artist had his pictorial imagination in such working order as to construct these pictures with no other material data than blueprints of plan and elevation of the various intricate forms — is simply beyond belief.

These drawings should be in a government museum of standards in a glass case, along with the platinum pound weight, yardstick, etc., to show the world and what comes after just what a mechanical drawing should be.

Not only that, but the rendering is a work of art, exact and life-like, and done with a delightful freedom of technique. They have a decided decorative quality, and in an appropriate setting could well serve as unique mural decorations.

I doubt if there are drawings anywhere that can in any way compare with these for perfection in showing what a stupendous piece of machinery is going to look like when finished.

The pity is that comparitively few will know about them, for their creation should be world news.

(signed) Maxfield Parrish

Cornish, New Hampshire, December 15, 1938. (Unpublished MS)

Chapter XVI

1. In the Visitors' Gallery of the Palomar Observatory there is a pictorial history of progress in telescope-making. Russell Porter constructed a wooden replica of the first reflecting telescope made in 1666, by the twenty-four year-old Isaac Newton. Compare this with the giant 200-inch reflector. (Unpublished MS)
2. a) In May 1949, *Popular Astronomy* published on pages 235-240 an obituary article entitled "Russell W. Porter 1871-1949," by the author. (Ed.)
b) In February 1951, *Popular Astronomy* published on pp. 79-87 an article entitled "The First Optical Tube (Astronomical Progress through the Centuries)", by the author. (Ed.)
c) A variety of articles for the *Springfield Reporter* and other newspapers. (Ed.)
d) In February, 1950, *The Astronomical Society of the Pacific* published in its Leaflet No. 251, an article entitled "Alhazen and the Telescope," by the author. (Ed.)
Following are two excerpts from his unpublished manuscript that relate to the preparation of the above article in the 1940's:

(1) Alhazen an Arab born in Basra, Iraq, around 960, died in Cairo, Egypt, 1058. He was the most noted astronomer, mathematician, and optician after Ptolemy. Originals of his writings were found in Oxford, Paris, and Leyden. Published in manuscript form by 1270 they were printed three hundred years later. Elmer H. Douglas, scholar of Arabic language and literature, examined some of the Alhazen material in the Paris libraries and shared his impressions with me.

(2) In Middleburgh, Holland, Loyal S. Wright obtained for me the following facsimile letter signed by the son of the man said to have made the first telescope:

In the year 1590, the first tubes have been made and invented by

Sacharias Jansen in the town of Middleburgh in Zeelant. The length, 15-16 inches has been used to look at the stars and the moon at night (wherein is much interesting material to be seen) . . . I am afraid I cannot give any further details.

Middleburgh 30 January 1655
Johannes Sachariassen

Notwithstanding the evidence furnished in the above document received through Loyal S. Wright, most authorities are inclined to give credit to Jan Lippershey for making the first telescope in 1609. (Unpublished MS)

3. The words "glacier-created wonder-land of Yosemite" remind us of John Muir whose life and writings the author admired. (Ed.)
4. *HAMLET* Act 5 sc. 2, 1. 10, from *Complete Works of William Shakespeare,* Vol. II page 631. Edited by W. G. Clark and Aldus Wright, Nelson Doubleday, Inc., New York.

Epilogue

1. In the spring of 1948, because of crippling arthritis, my father resigned from his responsibility in the Astrophysical Instrument Shop at Cal Tech. He and Mother left Pasadena in the fall of 1951 to live in Somerville, New Jersey, with my sister Candace and her husband Joseph B. Hartshorn. It was there, during our furlough from North Africa, that I first read portions of the manuscript of my father's autobiography and witnessed the suffering he endured with diminishing hope as he struggled to complete what he called *The Last Journeyman Machinist.* He died April 17, 1953. (Ed.)

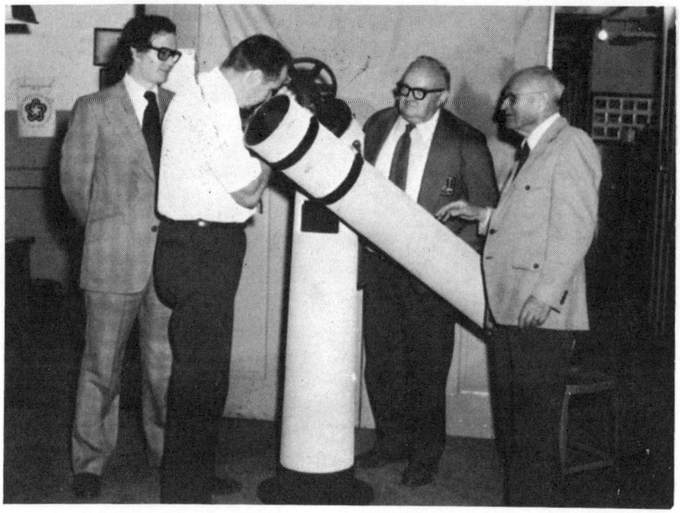

John Caldwell, Elliot Mattison and Fay Kingsbury, in charge of the restoration of "Antares." At right, Ernest Flanders viewing the telescope that OSM built. Courtesy Ernest V. Flanders.

INDEX

Note: Bold page numbers indicate pictures.